ALL GAVE SOME
SOME GAVE ALL

NORTH GEORGIA COLLEGE
KILLED IN ACTION: VIETNAM

PROVIDED BY
ATLANTA VIETNAM VETERANS
BUSINESS ASSOCIATION

WWW.AVVBA.ORG

www.deedspublishing.com

All pictures, unless otherwise noted, are from the collection of North Georgia College and State University.

Cover design and layout by Mark Babcock

Cover photo by Bob Babcock

Printed in the United States of America
Published by Deeds Publishing, Marietta, GA

First Edition, 2010

For information write Deeds Publishing, PO Box 682212, Marietta, GA 30068 or www.deedspublishing.com

ISBN 978-0-9826180-2-8

CONTENTS

DEDICATION

To all who have served in the United States military -
all gave some, some gave all.

Acknowledgements

Information for the honorees detailed in this work came from many sources. Students attending North Georgia College & State University in the fall of 2009 provided a significant portion of the material for North Georgia's Vietnam Honorees. Eight students, under the direction of Dr. Eugene Van Sickle, conducted research, personal interviews with friends and family of the honorees, and worked with alumni relations staff and with history faculty at North Georgia College & State University to compile as accurate a profile as possible for each individual. Based on that research, each student began the writing process, producing drafts on the lives of 26 of the 27 men. Without their hard work, this book would not be possible. The students who helped produce this book are: Matthew Bennett, Brenna Buchanan, Clay Comer, Adam Farrar, David Nichols, Peter Rehm, Daniel Sarratt, and Jordan Whiting.

NGC graduates Tony Faiia, Carl "Skip" Bell, and Jim Ruska, along with Dr. Eugene Van Sickle, meticulously read each draft of this book, always seeking out more information to honor those NGC students who made the ultimate sacrifice in Vietnam.

Members of Atlanta Vietnam Veterans Business Association (AVVBA), led by Memorial Committee Chairman, John Absalon, committed untold hours to insure the program to honor these 27 veterans was accomplished in the most professional manner possible – this book is one part of that overall effort.

Bob Babcock, AVVBA member and founder of Deeds Publishing, donated his time and talents to lead this project and make this book a reality. His partner, Mark Babcock, provided the editing, design, and layout work to insure this book is as good as it can possibly be.

For all who contributed in any way to the completion of this permanent tribute to these fallen NGC students, we thank you.

Letter From The President of AVVBA

April 2010

Since 1987, the Atlanta Vietnam Veterans Business Association (AVVBA) has held a memorial ceremony each year in the Greater Atlanta Area to honor Atlanta area men who lost their lives in Vietnam. Most of our honorees have been individuals, although in the year 2000 we honored, at the Atlanta History Center, all Atlantans killed in Vietnam.

Our objective with these memorials is to honor the sacrifice of these individuals and their families. Our focus has been on the families and we have attempted to convey to them that we as veterans and we as a larger society still owe them the appreciation and recognition many were denied at the time of the death of their loved ones.

With the current conflicts in Iraq and Afghanistan and other places around the world, this need to right the wrong of what happened in America in the Vietnam era has become even more important. In these recent conflicts, our country has, for the most part, rallied around our fighting men and women, honored them, expressed our gratitude, and supported them both while on the field of battle and when they returned home, with and without injury, alive and deceased.

Several current servicemen have attributed this groundswell of public support to the fact that our nation learned from the mistakes we made in the Vietnam era, and will not allow our warriors to be treated disrespectfully as they were during that time. The AVVBA continues to honor those who died in Vietnam as a reminder and example to the larger society that the fallen warrior must be honored and the sacrifice of his family must be recognized and appreciated and validated.

This book is dedicated to those 27 North Georgia College students who paid the ultimate sacrifice in Vietnam. They will never be forgotten.

Sincerely,

Alan C. Gravel
President, AVVBA

Letter From The President of NGCSU

April 2010

As one of only six senior military colleges in the United States, thousands of North Georgia alumni have served in military assignments – in times of peace and conflict – where they distinguish themselves through their demonstration of the university's core values of courage, integrity, loyalty, respect, service, truth, and wisdom.

Through these values, North Georgia provides an environment of academic excellence that has prepared generations of students to be professional, civic, and military leaders who have had a positive impact in their communities and around the world.

In the Vietnam War, 27 North Georgia alumni lost their lives in service to their country. When they were killed, their tours of duty in Vietnam ranged from as few as 26 days to 317 days into a second tour of duty, and their ages ranged from 21 to 42. They all left behind friends and family members who know this sacrifice first-hand. They also left behind a university that is proud of their service and remembers them with gratitude and respect.

In partnership with the Atlanta Vietnam Veterans Business Association, we are privileged to honor these men and their families during the university's 2010 Parents-Alumni Weekend. Thank you for helping us pay tribute to this deserving group of men.

David L. Potter
President, North Georgia College & State University

Introduction

At a debriefing meeting following the dedication of their annual Vietnam memorial in 2008, members of Atlanta Vietnam Veterans Business Association (AVVBA) started discussing a way to honor the 27 students from North Georgia College (now North Georgia College & State University) who had lost their lives in Vietnam. Several NGC graduates who are AVVBA members, and others, eagerly pursued the idea. Thoughts ranged from a traditional stone monument at North Georgia to a book honoring the service and sacrifice of those men.

Since a stone monument to fallen students was already in place at North Georgia, adding another monument was not pursued, but the idea was not dropped. After the annual monument dedication in 2009, it was decided to make a ceremony at North Georgia the primary focus of AVVBA's memorial efforts in 2010. Contacts were established with the university and its ROTC department to begin the planning process. It was determined that the best time to hold the memorial service would be at the annual Parents-Alumni Weekend, with the ceremony to be held on the drill field of North Georgia College & State University on April 17, 2010.

In lieu of a stone monument, focus was placed on producing this book to provide a permanent record of those 27 NGC students who made the ultimate sacrifice. Families and friends were contacted, memories were written down, and this book began to take shape. Although the chapters on some of the honorees are shorter than others, the tribute is equal to all who made the ultimate sacrifice in service to our country.

Current and future students, family, friends, and others can now forever read about those who came before them - men who "more than self their country loved."

During the Vietnam War, the North Georgia College & State University Corps of Cadets had an approximate personnel strength of 500 men. Twenty-seven former members of the Corps of Cadets were killed in action during that war.

They served in the U. S. Army, the U. S. Marine Corps, and the U. S. Air Force.

They represented the best of their generation. During a tumultuous time in the history of our Nation, they demonstrated their patriotism and valor by serving in a war that was questioned by many of their contemporaries.

The purpose of this memorial ceremony is to honor the sacrifices made by these former Cadets, and also the sacrifices made by their families.

DEDICATED BY
THE ATLANTA VIETNAM VETERANS
BUSINESS ASSOCIATION
17 APRIL 2010
ON BEHALF OF ALL WHO
SERVED THEIR COUNTRY

MEMORIAL DEDICATION
IN HONOR OF
27 FORMER NORTH GEORGIA CADETS
KILLED IN ACTION IN THE VIETNAM WAR

Patriotic Music Selections	Golden Eagle Band
Sounding of 1100 Hours	Golden Eagle Band
Flyover	94th Airlift Wing, Dobbins Air Reserve Base
Presentation of the Colors	Cadet Color Guard
National Anthem	Patriot Choir
Invocation	Cadet Chaplain C/CPT Zacharie Dumont
Opening Remarks	Alan Gravel, President, AVVBA
Welcoming Remarks	David Potter, Ph.D., President, NGCSU
Introduction of Speakers	Bruce Holroyd, Chairman, AVVBA
Keynote Speaker	MG James E. Livingston, USMC
Reading of the Honorees' Names	KIA Honor Readers
Folding of The Flag Ceremony	Wayne Witter, AVVBA
Presentation of The Flag	MG James E. Livingston, USMC
"Mansions of the Lord"	Patriot Choir
Benediction	Cadet Chaplain C/MSG John Kishimoto
Service Songs Medley	Golden Eagle Band
Vietnam Era Flyover	Army Aviation Heritage Foundation
Closing Remarks	Alan Gravel, President, AVVBA
Retirement of the Colors	Cadet Color Guard

Ceremonial Firing to Honor the Fallen	Blue Ridge Rifles
Silver Taps	Golden Eagle Band
Huey Medevac Flyover	Army Aviation Heritage Foundation
Retire to Reception	Alan Gravel, President, AVVBA

If you would like to add an online tribute to any of these veterans honored here, or any of the other 58,000+ whose names are on the Vietnam Memorial Wall, you can do that at WWW.VIRTUALWALL.ORG.

Those NGC Students Who Made the Ultimate Sacrifice in Vietnam

Earle John Bemis
Class of 1968

Name: Earle John Bemis

Born: December 2, 1945

Hometown of Record: Marietta, Georgia

High School: Sprayberry High School

NGC Years: 1963-1968

Date of Death: June 1, 1969

He is honored on Panel 23W, Row 35 of the Vietnam Veterans Memorial.

Born in London, England on December 2, 1945 to Earle A. Bemis Jr. and Florence Alice Taylor (now known to her friends as Florence Phinney), Earl John Bemis grew up the oldest child in a close knit family. His father served in the United States Army Air Corps during World War II. While stationed in London, he met Florence. The two were married at St James' Church in London. Eleven months later, Earle was born in Guy's Hospital. While living in England during the early to mid 1940's, the Bemis family experienced firsthand the hardships and tragedy of World War II. As the war in Europe came to a close, SGT Bemis received his orders to move back to the United States. Wanting to keep the family close, Mrs. Bemis and her son left her family and sisters to make the long trip across the Atlantic for a new life in Jersey City, NJ, where her husband began a career in engineering with Western Electric. Five years later, Earl became a United States citizen.

Earl grew up attending North Arlington Elementary School, and as a young boy, enjoyed playing in the band and swimming. Although he had to study much harder than his younger sister, Beverly, Earle was a good student and a passionate young man - a trait that distinguished him for the rest of his life. In 1959, Earl started at North Arlington High School, but his father was asked to transfer to the new facility in Dunwoody, GA, so once again the family moved. They took up residence in Marietta, GA where Earle continued his high school education at Sprayberry High, and participated on the track team and his school marching band.

Earle graduated from Sprayberry High School in '63 and enrolled for the fall quarter at North Georgia College, where he declared his major as Junior High Education, with a specialization in Biology.

Earle's passion for life and people made him instantly well liked and respected among his comrades in Band Company, where he served for two years in the Third Platoon. His classmate and close college friend, Warren Kirbo, states, "At our first meeting, Earle Bemis seemed to be one of those guys 'central casting' would have cast as an 'extra.' A good looking guy, not too tall and slightly built. He seemed quite satisfied with whatever hand he was dealt."

The members of Band Company were remarkably close and spent much time together. Kirbo recalls, "Band Company was like a fraternity, and it was the only company on campus that had a dorm to itself." Kirbo still remembers the times he served with Earl protecting the company Christmas tree from the Rifle Company and Lewis Hall.

During his sophomore year in 1965, Earle joined the Order of Colombo, a unit founded in 1962 by Louis P. Colombo, specializing in mountain warfare and survival tactics. Earle's participation in the Order of Colombo was not surprising, as he was an avid outdoorsman. He also joined the John C. Sirmons Chapter of the Student N.E.A, an organization for students whose vocational interests was teaching. In 1965-66, significant changes in the structure of the Corps of Cadets took him away from Band Company to Bravo Company.

Throughout his college career, Earle stayed close to his younger sister Beverly. Her fondest memories of Earle are the moments he took to spend with her whenever he could. He took the time to teach her to rappel and to conquer fear by exploring the natural, untouched caves in the North Georgia and Alabama mountains. "He was an excellent big brother, always looking out for me and making an effort to teach me where he could, discipline me when he needed to, sharing so much of his time as if he knew it may not be forever." she recalls.

In 1967, Earle started his student teaching at J.J. Daniels Middle School in Marietta. His passion for teaching and dedication to his vocation earned him the trust and respect of his students and teachers. He always making himself available to his students when they needed help.

Earle graduated from North Georgia College in 1967 with a B.S. in Junior High Education, and planned to continue his passion to help others by serving in the Peace Corps, but at his father's urging, took a commission in the United States Army. This decision would plague Earle's father for the rest of his life.

Shortly after his commissioning into the Army, Lt. Bemis' took up his station at Ft. Bragg and then Ft. Lewis, where he earned a silver tray for his outstanding service. He began his tour in Vietnam on April 15, 1969, as an infantry platoon leader. Lt. Bemis' letters home indicated that he took the horrors of war that were inflicted on the Vietnamese civilian population to heart. In one letter he spoke of how he had befriended two Vietnamese children whose parents had been killed. He was teaching them English and they were teaching him to speak Vietnamese.

On June 1, 1969, tragedy struck the Bemis family. A U.S. helicopter went down near the zone in which Lt. Bemis' platoon was engaged in combat. The young lieutenant stormed the chopper and pulled two men to safety. Pam Rossett narrates the event as follows, "The fact is Earl sacrificed his life while saving the life of two others. When a chopper was shot down, under fire Earl went out and pulled two crew members to safety. He took a bullet… while trying to rescue a third."

The twenty-three year old lieutenant's body was recovered and sent back to Marietta, GA where his funeral arrangements were held at Faith Lutheran Church with the Rev. Edgar A. Trinklein officiating. An article printed shortly after Lieutenant Bemis' death revealed that he was engaged to be married to Sylvia Edwards of Marietta, possibly on his rest and recuperation leave to Hawaii scheduled for September 1969. He was posthumously awarded the Purple Heart and Silver Star for his service in Vietnam.

Lieutenant Earle John Bemis spent his short life in the service of others, and those who walk into the Faith Lutheran Church Sanctuary will continue to be overlooked by his warm, compassionate presence, as a large memorial cross hangs in the memory of him. It stands as an everlasting testament to a sincere and genuine person who devoted his life to making this world a better place.

Special thanks to Florence Phinney, Beverly Bruce, and Warren Kirbo for their contribution to this biography.

Burton A. Blanton
Attended 1958-1959

Name: Burton A. Blanton

Born: September 30, 1937

Hometown of Record: Charleston, South Carolina

High School: unknown

NGC years: 1958-1959

Date of Death: March 30, 1966

He is honored on Panel 6E, Row 62 of the Vietnam Veterans Memorial.

Burton A. Blanton, or as his friends called him "Skip" Blanton, was born on September 30, 1937 in Charleston, SC. He attended North Georgia in 1958-1959 and was later commissioned into the United States Army. He was commissioned as a helicopter pilot and was part of flight class 61-3 and flew the UH-1D. He made many friends during his time in service because of his great personality and his love of poker.

LTC David Lacromb said of him, "In 1962-63 I was assigned to the 8305 ACR Company, Ft. Rucker, Alabama. That is when I met Skip Blanton. I knew him for several months before I was reassigned to Hawaii. He stayed in 8305th until it became a test unit for the 1/9th Cavalry, 1st Cavalry Division at Ft. Benning, GA. When the new unit was judged to be combat ready in August 1965, the whole Division went to Vietnam.

"Skip went to Vietnam along with that unit. He was assigned to "A" Troop 1/9 Cavalry Squadron, 1st Cavalry Division. My unit (170th Aviation Co) went to Pleiku, Vietnam in November 1965.

"When I knew Skip, he was a reconnaissance pilot training on the H-13 E model Bell Helicopter. When he went to Vietnam, I think he flew the UH-1D model Huey. Since I knew many of the pilots in "A" Troop, I had the occasion to visit them at their headquarters located at An Khe while I was assigned to the 170th in Pleiku. I had the opportunity to meet with Skip and several other friends in early March 1966.

"Skip was killed on March 30, 1966, while on a combat mission near the Laos Border near the Chu Pong Massif. I believe he was killed flying a Huey to transport a new commander and others on a combat mission in the Ia Drang Valley, near the Chu Pong. Skip was married and was one of the nicest friends a person could have. He was a dedicated soldier and aviator. He was a good pilot, and I am proud to be a friend of his."

Skip Blanton spent six years in the United States Army and rose to the rank of Captain. He was killed by enemy fire while doing a mission in South Vietnam. He is survived by his wife.

WILLIAM A. BRANCH
CLASS OF 1963

Name: William Anderson Branch

Born: July 11, 1941

Hometown of Record: Fitzgerald, Georgia

High School: Georgia Military Academy

NGC years: 1959 – 1963

Date of Death: June 6, 1970

He is honored on Panel 9W, Row 18 of the Vietnam Veterans Memorial.

General Dwight D. Eisenhower said, "Pull the string, and it will follow wherever you wish. Push it and it will go nowhere at all." Captain William "Bill" Anderson Branch understood the values of good leadership and exemplified outstanding military leadership. Captain Bill Branch knew his strengths and utilized them to be the best leader possible. From his austere beginnings in Fitzgerald, Georgia, to the ultimate price he paid in 1970, Bill Branch lived life.

Bill Branch was born on July 7, 1941. His family is originally from Fitzgerald, Georgia. While Bill attended North Georgia College, his family, Rusty and Margaret Branch, moved to Belleville, New Jersey. His mother has since returned to Fitzgerald, Georgia. As Bill Branch aged, his military aspirations blossomed. He attended the Georgia Military Academy located in Atlanta, Georgia. Today, this school is known as Woodward Academy. The school had a strong military tradition until 1966 when it changed its name.

Bill Branch was an only child. He married Judith Ann Dunn of Decatur, Georgia, also a graduate of the NGC class of 1963. Before his death in 1970, they had a daughter, Jennifer. She is now Jennifer Branch Denard and has three children. Bill Anderson Branch's grandchildren are Elijah Branch Denard, Owen Anderson Denard, and Emma Grace Denard. He is also survived by his mother and his wife, who has remarried.

Benjamin Franklin said, "An investment in knowledge always pays the best interest." Bill Branch bought into this ideal. His four years at North Georgia College were well spent. He studied History and earned a B.A. in History in 1963. During his college tenure, he helped found the Mountain Order of Colombo. This military organization developed as a result of a demonstration that caused North Georgia College students to want a special mountaineering unit.

Bill Branch joined the ranks of the National Society of the Scabbard and Blade and held the rank of major while on campus. He served in the Aggressor's Platoon at Summer Camp.

He also participated both in the Blue Ridge Rifles and the Non-Commissioned Officer's Club. His non-military activities included the Dramatics Club, Y.M.C.A, and his tenure as the Vice-President of the Junior Class. To place a capstone on all these activities, he was elected as the Most Versatile during his senior year. All these achievements can be found in the 1963 *Cyclops*.

Bill Branch understood leadership. He was a founding member of an elite military unit on campus, he provided leadership in the civilian campus organization, and he was an Executive Officer on the First Battle Group. Bill was well-liked by the college community.

According to his daughter, his college friends included GB Collins, Henry Camp, Mitch Mitchiner, and Ed Scholes. More friends found his profile on the Vietnam Veteran's Memorial Fund and posted memorials. These included Dr. John C. House, Rusty Hightower, and James L. Newborn. Reading through these citations, one recognizes that Branch was a well respected student at North Georgia College.

Recalling Benjamin Franklin's statement, Bill Branch took what he knew and enlisted in the Army in 1963. His military service included two tours of combat duty. The first was from 1966 to 1967 when he served as a MACV (Military Assistance Command Vietnam) advisor with the Second Battalion, 46th Army of the Republic of Vietnam (ARVN) Infantry in the Long An Province. The second tour lasted from 1969-1970 as Alpha company commander of the 2/14th Infantry, 25th Infantry Division.

Later, Captain Branch moved to Headquarters and Headquarter Company (HHC) as a Tactical Staff Intelligence Officer before he was killed on June 6th, 1970. According to official records, Captain Branch died during a helicopter reconnaissance mission. According to Jeff Fuller, who served with Captain Branch, his helicopter was shot out of the air by hostile forces with small arms fire. The helicopter missions provided reconnaissance for the ground units. Captain Branch excelled at this type of mission.

The aforementioned James L. Newborn notes that Captain Branch exhibited an artistic ability. This detail is confirmed by a memorial site that his daughter posted. Her site says, "They call him 'the good captain.' They tell me that he cared about his men, that his maps were detailed and amazing. That his artist's eye helped gather the intelligence others failed to see. Many say he's responsible for getting them home." Even this citation speaks to the influence that this hero had on his fellow soldiers.

Army biographical researcher Clay Marston states that Captain Branch posthumously received the Silver Star, Bronze Star with 'V' for Valor device, Purple Heart, National Defense Service Medal, Vietnam Service Medal, Republic of Vietnam Campaign Service Medal, and was entitled to wear the Combat Infantryman's Badge." Speaking with his daughter, she says that he actually received two Purple Hearts.

Captain William Anderson Branch embodied the principles of goodness. The records all point to a man who was quick to inspire. His memory lives on in the lives of those that he has touched. From his college friends who recall his free spirit and innocence, to the soldiers who lived through hell with him, and on to his daughter who just wanted to know how the man lived. Looking at the evidence, Bill Branch lived a life that was heroic. General Eisenhower would have said that Captain Branch did a lot of pulling. In other words, he was willing to lead by example. This willingness to lead through actions rather than mere commands transforms the man into a hero.

Captain Branch and his wife, Judith Ann Dunn
Pictures contributed by Jennifer Branch Denard

William and Judith Branch, with their daughter
Jennifer

WELBORN A. (BILL) CALLAHAN, JR.
CLASS OF 1965

Name: Welborn A. (Bill) Callahan, Jr.

Born: March 6, 1943

Hometown of Record: Columbus, Georgia

High School: Baker High School

NGC years: 1961 - 1965

Date of Death: March 3, 1967

He is honored on Panel 16E, Row 7 of the Vietnam Veterans Memorial.

Welborn A. (Bill) Callahan, Jr. was born on March 6, 1943 in Columbus, GA and lived on Ft. Benning. Callahan's father, Welborn Callahan Sr., was Post Sergeant Major at Ft. Benning, so Callahan grew up as an "army brat". He graduated in 1961 from Baker High School. As a senior in High School, Callahan was a member of the Senior Student Council Representatives and Officers and was also a member his Junior year. Callahan was also a member of the "B" club at Baker High School during his Junior and Senior years, whose mission was "furthering high ideals of sportsmanship and arousing the schools interest in sports." In addition to membership in the "B" club, Callahan also played on the High School football team as #51. He was elected as Junior and Senior Class Treasurer. In the yearbook's Senior Directory, Callahan is listed as a Jr. Civitan his Junior and Senior years, a Who's Who student his Senior year, and credited with the superlative of "Cutest." He was also a member of the Junior Red Cross his Freshman year, Varsity Football his Sophomore, Junior, and Senior years, Varsity Baseball his Freshman, Junior, and Senior years, and ROTC his Freshman year. In his senior photo in the Arrowhead yearbook Callahan leaves the caption, "I never get lost, someone is always there telling me where to go."

The alumni at the NGC reunion of the class of '63 remembered Callahan with great esteem. Classmates described Callahan as a "great cadet" and as a "great guy." Maurice Healy went to high school with Callahan, a year ahead of him in school. Healy remembers that when he went to NGC his first year, Callahan "sneaked his girlfriend" from him while Callahan was still a senior in High school in Columbus. Portia McDonald remembered that she and her husband purchased a rug from Callahan's wife after he had passed. Finally, Billy York remembers playing football with Callahan while they were both in A Company.

Bill Callahan attended North Georgia College from 1961-1965. As a freshman he was a member of A Company, and played on the company football team as #40. In 1963 as a sophomore, he was a member of A Company's First Platoon and also a member of the aggressor's platoon. During this time he also became an officer of Sigma Theta Fraternity. In 1964, his Junior year, Callahan switched to Echo Company and became a member of the First Platoon. During this year he was also a member of the Non-Commissioned Officer's Club and the Baseball team. In 1965, his Senior year, Callahan switched once again and became a member of B Company, earning the position of Company Commander.

During his time at North Georgia College, Callahan served in a variety of different positions and a number of different organizations. These include All Star Football, Scabbard and Blade, Aggressor Platoon, Sigma Theta Fraternity, B.A. Club, NCO Club, Officers Club, Intramural Sports, Varsity Baseball, Outstanding Platoon Sergeant, and DMS. Callahan graduated North Georgia with a B.S. Degree in Business Administration.

Callahan's tour of Vietnam began on Oct 21, 1966. Callahan was a member of the Regular Army and obtained the rank of 1st Lieutenant. Callahan served in C Company, 2nd Battalion, 503rd Infantry Regiment, 173rd Airborne Brigade. The Brigade was the first major ground combat unit to serve in Vietnam. The Brigade participated in the first and only combat air jump in Vietnam, on February 22, 1967, during Operation Junction City. The Brigade was also the first unit to set foot into War Zone D, where they destroyed enemy camps, and also helped introduce small long-range patrols.

Bill Callahan was killed in action in Tay Ninh Province, South Vietnam on March 3, 1967, during an air assault. Callahan was killed by a sniper before he hit the ground leaving the helicopter. He was survived by his wife, Mrs. LaJuan Callahan. He is buried in Ft. Benning Post Cemetery. His name is on Panel 16E - Row 7 on the Vietnam Memorial Wall in Washington D.C.

Callahan was a remarkable man and is still remembered by his classmates and his friends. On the North Georgia College Reunion website, an old friend and a fellow veteran reminisce about him. John T. Radney, his best friend from Columbus, GA writes, "We were little boys together. We fought, dated the same girls when we got older, played sports, on the same side and against each other, and were always together. We got into some real messes, too. My life and my success is a credit to him. I loved him."

John Shope, who got to know Callahan during their senior year at North Georgia, remembers a day during inspection for B Company in Barnes Hall. The Freshman Frogs had "hit huts" so hard when upperclassmen had come out into the hall that they actually left imprints of their bodies into the sheet rock. Since Callahan was company commander, he was the one who had to pay for the repairs. Shope remembers Callahan as very serious about the military and as a "quiet, reserved guy." He recalled that hundreds of enlisted people attended Callahan's funeral at Ft. Benning; probably acquaintances of Callahan's father.

Doug Imes, Callahan's roommate during their senior year, remembers an incident when he, Callahan, and others were doing exercise and Callahan fell off a bridge into the Chestatee River. Callahan was holding an automatic rifle between his legs. He did not want to let the rifle go because he was afraid he would have to pay for it. Imes and others feared that Callahan would drown if he did not let go of it. Eventually, Callahan did let go and drifted down river where Imes and a few other boys fished him out about fifty yards downstream. Imes also remembers seeing Callahan again when they were in Infantry school after they had both left NGC. Callahan, excited, told Imes that his father was instrumental in getting him an assignment to go to Vietnam.

Callahan received a Silver Star, the nation's third highest award for valor, the Purple Heart, Vietnam Service Medal and a National Defense Medal. Callahan's friends will always remember the time that they spent with him, and will always honor him for the sacrifice that he made for this country.

Special thank you to Billy York, John Shope, Doug Imes Robert Sage, Rusty, William Ethington, Maurice Healy, Nick Heldreth, and Portia McDonald for providing research on this honoree.

Ralph Durward Cordell
Class of 1957

Name: Ralph Durward Cordell

Born: July 3, 1935

Hometown of Record: Hartwell, Georgia

High School: Hartwell High

NGC years: 1953 – 1957

Date of Death: January 15, 1967

He is honored on Panel 14E, Row 33 of the Vietnam Veterans Memorial.

Ralph D. Cordell was born in Hartwell, Georgia on July 3, 1935. Ralph lived in Hartwell his entire childhood and attended North Georgia College after high school. At North Georgia College, Ralph had many accomplishments. In his four years there, he received a B.S in business administration, became the Second Battalion Commander, and was a DMS (Distinguished Military Student). He was also in many organizations such as the Dramatics Club, Scabbard and Blade, the Officer's Club, the NCO Club, the BA club and he was the Vice President of Sigma Theta fraternity.

During Cordell's senior year, he married Bobbie Cordell. Unfortunately, at the time it was against school rules to be married and have a command position so later that year he had to step down.

In Vietnam, Cordell was assigned to the 1st Logistical Command, who oversaw the delivery of supplies for the US Army in Vietnam. Cordell died on January 15, 1967 when the helicopter he was traveling in suffered a mechanical failure and crashed near Can Tho. As a Major, he was overseeing a construction project for the Special Forces near the village. Shortly after takeoff, the aft blade failed, sending the Chinook crashing into a rice paddy. One of the first people on the crash site was Terry Gordy, one of his classmates at North Georgia College. He spoke about how disturbing the crash was with the craft nearly split in half. Cordell was just a few months shy of ending his tour in Vietnam.

Today, Cordell's family is living in North Carolina, and one of his sons continued the family military tradition by attending West Point.

Officers of the Sigma Theta Fraternity, Cordell is second from the left. (Photograph found in the NGC Bugler)

William Carroll Elrod, Jr
Class of 1962

Name: William "Stick" Carroll Elrod, Jr.

Date of Birth: September 8, 1940

Hometown of Record: Byronville, GA

High School: Georgia Military Academy

NGC years: 1958 – 1962

Date of Death: April 14, 1971

He is honored on Panel 4W, Row 123 of the Vietnam Veterans Memorial.

William Elrod, Jr. was born on September 8, 1940, to William Carroll Elrod, Sr. and May Kittles Elrod. He grew up in Byronville, Georgia. He attended school in Dooly County until the 7th grade, after which he began attending Georgia Military Academy for high school. The Georgia Military Academy was located in College Park, Georgia. His time and education were intended to prepare him for a career in the armed services. After graduating in 1958, he chose to attend Georgia's military college, North Georgia College. He entered in the fall of 1958. During his time in school, he was an English major. Elrod succeeded in languages as well. He was fluent in French, German, and Vietnamese. Coinciding with his degree, he worked as an editor/author for the school's newspaper, the *Cadet Bugler* in 1960 and 1961. He wrote in the military portion of the paper. He also sponsored the lovely ladies in Phi Omicron.

He achieved well in the school's military, and quickly climbed through the ranks. As a senior, he roomed with his close friend, Conrad Easley. Easley served as the Commanding Officer of Alpha. By his senior year, he was a first Lieutenant in the First Battle Group. He was also the Executive Officer for his company, Company Alfa. Easley described Elrod as a "great second in command." He also earned a few significant achievements during his time at North Georgia College. His squad and his battalion were both awarded "Best Drilled". Sham battles were fought to ensure success in the upcoming Summer Camp. Seniors and Juniors competed against one another. Elrod and Easley were both Juniors at the time. The Seniors held an ammo dump near Crown Mountain. The only way the Juniors could approach the ammo dump was by crossing the Chestatee River. Easley described Elrod as the "skinniest guy he knew in his life." Unfortunately for Elrod, the river had risen dramatically. Looking back, Easley noticed Elrod missing. Glancing downstream Easley noticed Elrod holding onto a tree for his life. Elrod kept his grip and was rescued in fifteen mintues. They managed to capture the ammo dump, but were miserable the rest of the night. Perhaps thanks to the experience they shared together at Crown Mountain during the Summer Camp of 1961, in a field of thirty-four different colleges, North Georgia prevailed to stand above the rest. Thanks to the success of Easley and Elrod in Summer Camp, they received higher ranks than they previously would have. Elrod was an excellent cadet, to be sure. Due to Elrod's success in school and in the corps, Easley described him as "Sharp as a tack."

Thanks to his rather limber and skinny build, Elrod earned the nickname "Stick" amongst his classmates. He was also known for having a good sense of humor. Despite his humor, he took his time in the corps of cadets seriously and was described on one occasion as being a "hard-ass." Nevertheless, he was always looking out for others and trying to do what was right. His roommate during his years at college was his close friend Conrad Easley. He met his wife at North Georgia College in 1959, and her name was Claudia Kelly. Her fellow classmates described her as beautiful; Elrod was a lucky man indeed. They were married over winter break in 1961 and were the first to do so in their graduating class. Soon, they had their first child, a son named William Carroll Elrod, III. After graduating, Elrod was stationed in Fairbanks, Alaska, from 1963-1965. Conrad Easley, however, went onto medical school. During this time, his second child was born, a daughter by the name of Anne Lynn Elrod (Galloway). His third, Claudia Allison Elrod (Miller), was born in May of 1969. Allison, in particular, looked just like her father.

From August 1965 to May 1967, he served on the Infantry Board. During that time he was a major editor for the *Infantryman's Handbook*. His first tour in Vietnam was from May 1967 to May 1968. During his first tour, he served as an Advisor to South Vietnamese troops, as America had yet to formally commit troops. A talent for languages was primary the reason for his selection to join the conflict as an Advisor. By the time of his second tour, which began on August 7, 1970, he still served as an Advisor to South Vietnamese troops. He was assigned as a Senior Advisor, 3rd Battalion, 41st regiment, Army of the Republic of Vietnam. By this time, he had risen to the rank of Major in the army and served as a MACV Advisor. His final battle was to be fought in southern Vietnam at Kontum, April 13, 1971. The mission for him and his troops was to defend their position against the Viet Cong. As their position began to be overrun by their enemies, Elrod called to be evacuated. The helicopters arrived to evacuate the troops, but there simply was not enough room for everyone. Elrod opted to stay behind because he thought of his men first. Nightfall was quickly approaching, and regrettably for Elrod and his fellow soldiers, the helicopters were not allowed to fly at night. The helicopters were to return at first light the next day.

Unfortunately, The Viet Cong counter attacked. They overran the men who were left behind and took no prisoners. When the helicopters returned in the morning they found all the men dead. The Evac crew found Elrod bayoneted to death on April 14, 1971. George Whitely described him as a "real American hero, who gave his life for his men and for his country." For his sacrifice, he was awarded the Silver Star. Veterans of the Vietnam War, such as Whitely and Easley, are thankful to hear that their fallen comrades are being honored, especially after previous decades of neglect.

A special thank you goes out to George Whitley, Anne Lynn Galloway, Claudia Allison Miller, and Billy York for their contributions to Elrod's tribute.

ROBERT W. GARTH
CLASS OF 1961

Name: Robert W. Garth

Born: February 28, 1939

Hometown of Record: Madison, GA

High School: Madison High School

NGC years: 1958-1961

Date of Death: September 23, 1966

He is honored on Panel 11E, Row 2 of the Vietnam Veterans Memorial.

Bobby Garth was born in Morgan County, Georgia on February 28, 1939 to Robert W. Garth Sr. and Mildred Garth. Bobby had one brother and three sisters, and the family operated a dairy farm in Morgan County. Growing up on the dairy farm, he developed the work ethic, character, and dedication that insured he would succeed in every endeavor he attempted in life.

Even though he had to spend a great deal of time working on the farm, he was so well organized and motivated that he also had a very active high school career. He was very active in the FFA as well as a number of other clubs. He attended Boys State after his junior year and was an outstanding player for the Morgan County High School football team. In his senior year, his team won the Class B State Championship.

When he entered North Georgia College in the fall of 1957, his work ethic and advanced organizational skills gave him a huge advantage over most of his new classmates. He immediately established a reputation as the "go to guy" when one needed help. He was a tall, broad shouldered, handsome young cadet and was very popular with all the students on campus. He was assigned to Bravo Company in the Corps of Cadets and was a major factor in helping B Company win the Honor Company of the year award in his freshman year. He played intramural tackle football for B Company for four years.

Bobby had a great sense of humor and an easy going personality that endeared him to everyone on the campus. He was an outstanding cadet and a natural born leader. He had always wanted to be in the military, and as a child he played "war games" with homemade toys that he designed and constructed.

He loved horses, and at age 17 decided he was going to Texas to become a rodeo star. It didn't take long to discover that being a top rodeo performer required many years of work and practice. He also learned that it is very expensive to be in a rodeo unless you win some big purses. When he ran out of money, he returned to Georgia – that was probably the only activity that Bobby Garth ever attempted that was not successful.

In his sophomore year, he met Farrell Early and that relationship turned into a long term love affair that lasted the rest of his life.

At NGC he served as a Squad Leader, a staff NCO, and in his senior year, he realized his dream when he was selected to command Bravo Company. He earned a BA Degree in History, was on the Dean's List, a member of the NCO and Officer's Club, the Rex Fraternity, the Scabbard and Blade, and was selected for Who's Who in American Colleges and Universities.

He was anxious to join the active Army, so he worked hard and graduated in three and a half years and immediately entered active duty.

After graduating from NGC, he attended the Infantry Officer Basic Course and Jump School at Fort Benning, Georgia and then moved to his permanent assignment with the 101st Airborne Division at Fort Campbell, Kentucky. He was assigned as an Infantry Platoon Leader in E Company, 327th Infantry.

Bobby and Ferrell were married on August 27, 1961 and loved serving at Fort Campbell. In late 1962, Bobby was selected to attend The Army Aviation School at Fort Rucker, Alabama and then was assigned to fly in the 3rd Armored Division in Friedberg, Germany.

He began his tour in Vietnam on June 30, 1966 and was flying the O-1 (Birddog) with the 220th Aviation Company (Recon), 223d Aviation Battalion, 1st Aviation Brigade. He was stationed at Quang Ngai, VN.

On September 23, 1966, he was flying a recon mission with a non-rated Marine Corps Officer (Captain Frank H. Adams) in the back seat of his Birddog. During the mission, they received heavy ground fire and Bobby received a round through his right hip that traveled through the upper part of his body. Captain Adams reported that Bobby very calmly said, "Oh my God, I'm hit" and those were the last words he uttered.

Captain Adams, being non-rated, didn't know how to fly the aircraft and in fact, didn't have the controls to fly it. When an observer was in the back seat, the control stick was removed, and stowed, to give the observer more room for his maps. Therefore, Captain Adams had to retrieve the control stick from its stowed position and install it before he could control the plane. When he finally got the stick installed, the plane was only 100 feet above the ground. Fortunately, there were other Birddog pilots in the area that could give Captain Adams verbal instructions on how to land the plane. Again, unfortunately, Bobby's left leg had jammed the left rudder pedal all the way forward and Captain Adams could not compensate for it. When they touched down at the airfield, the plane was uncontrollable, and they crashed into a stack of live White Phosphorus Artillery rounds. Luckily, none of the rounds ignited.

Bobby was flown by marine helicopter to a hospital, but unfortunately it was too late. The next day, Captain Adams wrote Bobby's wife Farrell and told her, "He in essence, gave his (ultimate): his life, in a sad land, for love of family, God, and country."

Captain Blanton, an Assistant Professor of Military Science at NGC, notified Farrell of Bobby's death and assisted her through the next few days. Captain Howard Floyd, a close friend, escorted the body to Georgia and served as the Survivor's Assistance Officer. Both officers were very supportive of Farrell and the children.

During his career, Bobby was awarded the Distinguished Flying Cross, Purple Heart, VN Service Medal (2 Bronze Stars), Air Medal (3 OLC), National Defense Medal, VN Campaign Medal, VN Gallantry Cross with Palm, and the Republic of VN Order 5th Class.

Captain Robert W. Garth, a great soldier and man, a great husband, a great father, and a great friend was interred at Marietta National Cemetery in Marietta, Georgia.

Walter Murrah Gibson
Class of 1968

Name: Walter Murrah Gibson

Born: August 7, 1946

Hometown of Record: College Park, Georgia

High School: Arlington High School

NGC years: 1964 - 1968

Date of Death: October 28, 1969

He is honored on Panel 17W, Row 127 of the Vietnam Veterans Memorial.

Walter "Hoot" Gibson was born in Red Oak, Georgia on August 7, 1946 to Eddie and Rebecca Herren Gibson. Walter's father worked in the transportation business and his mother stayed home to raise Walter and his younger brother Glenn.

Walter grew up attending Atlanta Public Schools. "He was a pretty outgoing boy who had lots of friends," states Hoot's father. Walter received his secondary education at Arlington High School, where he graduated in 1964.

Hoot enrolled at North Georgia College in the Fall of 1964 and trained with Echo Company in the First Platoon. In 1965, Hoot was transferred to Delta Company where he trained with the Second Platoon before transferring to Golf Company, Second Platoon in 1966. In 1967, Walter served as a Platoon Commander in Charlie Company.

Hoot was remembered around campus for his brilliant sense of humor. He remained remarkably active throughout his time at North Georgia, serving as the Junior Class Treasurer and the chaplain for the NCO club. He graduated in 1968 with a B.S. in Business Administration.

After graduation, Walter immediately began his career in the United States Army as an Infantry Unit Commander in the 101st Airborne Division's 3rd Battalion, 506th Parachute Infantry Regiment, also known as the Currahees. He was stationed at Fort Polk, Louisiana for nine months before beginning his tour in Vietnam on July 18, 1969.

His commanding officer at Fort Polk, Jean R. Emery, writes, "I had nothing but the highest regard for this outstanding officer. He was extremely competent and well respected by all who knew him, both superior and subordinate. Upon his departure from this unit, I ranked him number one of ten lieutenants whom I rated."

Lieutenant Gibson lost his life in the morning hours of October 28, 1969 by Viet Cong small arms fire as he and members of his platoon were engaging an enemy position during a combat operation. He died instantly from his wounds. Captain Harry E. Rothmann writes, "Walter was one of the most outstanding young officers with whom I have been associated. He was a truly dedicated individual, whom we admired and respected. Walter was hardworking and conscientious in all that he did, and his personal courage on the battlefield won him the respect of all the officers and men in the Company."

Prior to death, Walter had been awarded the National Defense Service Medal, Vietnam Service Medal with One Bronze Star, Vietnam Campaign Ribbon, and the Expert Badge with automatic rifle bar.

Walter's body was recovered and taken back to Atlanta, GA where his funeral service was held at the Mary Branan Methodist Church. He is buried in the Westview Cemetery in section 79 B, Grave 3 Sermon on the Mount Section.

Hoot is remembered by those who knew him best for his brilliant sense of humor and excellence in military leadership. The smile of this young officer touched many lives who will be forever in debt for the character and fortitude they learned from Walter Murrah Gibson.

Special thanks to Eddie Gibson and Glenn Gibson for their contribution to this biography.

JOHN EDWARD GREENE
CLASS OF 1965

Name: John Edward Greene

Birth Date: September 4, 1943

Hometown of Record: Albany, Georgia

High School: Albany High School

NGC years: 1961 – 1965

Date of Death: March 13, 1972

He is honored on Panel 2W, Row 114 of the Vietnam Veterans Memorial.

John Edward Greene was born on September 4, 1943 in Albany, Georgia. John is survived by his brother, Mike Greene, who lives in Lilburn, GA, and his mother who resides in Birmingham, AL. After graduating from Albany High School in 1961, Greene decided that he wanted to attend college and picked North Georgia College. According to his brother Mike, NGC was an easy choice for Greene because he always desired a career in the military. It was at North Georgia that Greene would receive the education and experience that would shape him as a soldier.

He attended North Georgia College from 1961-1965. During his first three years of college, he was a member of D Company and played on the Company football team as #67. In 1965 when he was a senior, he switched to B Company and achieved the position of Third Platoon Leader. While attending North Georgia, Greene was a member of The Officers Club, the NCO club, the Blue Ridge Rifles, and the Best Drilled Squad. Greene also obtained the rank of Second Lieutenant and worked with the YMCA. In 1965, he graduated with a B.A. Degree in History and was a recipient of the Military History Award.

A few of Greene's classmates remember going to school with him. Billy York, who played football with Greene in D Company, remembers him as a "real, decent guy" who was "nice" and quiet". John Shope, who was the executive office of Bravo Company his senior year, graduated the same year as Greene and remembers him as a "very quiet, very inward" individual. Greene's younger brother, Mike, posted on the Vietnam Veterans Memorial Fund Website, "Johnny was my older brother. A great guy dedicated to the military service of his country." Mike Greene remembers his brother as an "introvert" and recalls him coming home from North Georgia during breaks and calling home from school.

Greene was commissioned upon his graduation from North Georgia, and his brother remembers attending the ceremony. Greene also took the bar exam and was accepted to, but did not attend, Tulane University. He also took a few law classes at the

University of Virginia. During his time in the service, Greene spent time at Fort Knox, Fort McPherson, and Fort Benning.

Greene began his tour in Vietnam on November 26, 1971 as a Captain in the Regular Army. He served with the MACV (Military Assistance Command Vietnam) Advisors, Headquarters Advisory Team 28. He was classified as an Infantry Unit Commander and a member of the General Staff.

Greene was killed at the age of 28 in a hostile ground encounter in Phu Yen Province, South Vietnam on March 13, 1972. The Army determined a gun or small arms fire caused his death. Greene's name is located on Panel 02W - Row 114 on the Vietnam Memorial Wall in Washington, D.C.

Greene greatly benefited his classmates through his participation in the Corp of Cadets at North Georgia and also with his easy going personality. He gave the ultimate sacrifice for his country in Vietnam. He will never be forgotten by his friends and family. Greene received the Purple Heart, a Vietnam Service Medal, and a National Defense Medal.

Special thank you to Mike Greene, Bill Burns, Billy York, Daniel Williams and John Shope for providing information in the research on this honoree.

RICHARD A. GWINN
CLASS OF 1968

Name: Richard A. Gwinn

Born: December 24, 1947

Hometown of Record: Miami, Florida

High School: Anchorage High School

NGC years: 1964 - 1968

Date of Death: September 26, 1969

He is honored on Panel 17W, Row 10 of the Vietnam Veterans Memorial.

Born in Quincy, Florida on December 24, 1947, Richard A. Gwinn grew up the younger of two brothers. His father served most of his life in the United States Army – a career that forced him to move his family around the world.

From 1950 to 1954, the Gwinn family lived in Dachau, Germany before moving to Ft. Bragg, North Carolina where they stayed for a year or two. In 1961, Mack's job led the family to Fort Richardson in Anchorage, Alaska where Richard attended and graduated from Anchorage High School in 1965. Mack W. Gwinn Jr., Richard's older brother, remembers Richard as being "a big-strong, smart kid who wasn't afraid of hard work."

Upon graduating high school in 1964, Richard enrolled in North Georgia College where his father held a teaching position in Military Science. While at North Georgia, Richard became close friends with Tom McLaughlin, or "Buddha" as he was called by his closest friends. Buddha thought much of Richard and spoke of him as being a quiet individual who never bragged on himself. "Richard was always out to help others. He was a great listener who would never let you get down. He took everything with a smile," states Buddha.

In his freshman year, Richard trained with Foxtrot Company in the Third Platoon before being transferred to the Third Platoon in Charlie Company in 1967. Also in 1967, Richard earned his initiation into Scabbard and Blade, a joint service honor society that unites cadets and midshipmen from all over the country in military excellence.

Buddha recalls that during Ranger School, Richard pushed everyone around him to be their best. He remembers that there were two Navy seals in their squad and the training "was rougher on them physically than Richard." Buddha states, "he [Richard] did everything possible to assist the Navy's best to obtain the prized Ranger Tab as he did all, including me." Richard and Buddha were the first to finish their patrols during tactics training where they went 3-0. Buddha attributes their success to the leadership and excellent map reading skills Richard possessed. "He was the best map reader I ever met," states Buddha.

Richard matched his excellence in military leadership with a remarkable performance in the classroom. All of Richard's classmates recall him as being one of the brightest, if not the brightest, individual in their class. Classmate and friend Warren Kirbo recalls that Richard earned a Rhodes Scholarship; however, Buddha said that "serving his country meant more to Richard than his Rhodes Scholarship."

Upon graduation from North Georgia College in 1967, Richard was commissioned into the Army as an Infantry Unit Commander in the Third Brigade, 82nd Airborne Division. Lieutenant Gwinn went to Vietnam in April of 1969, spending much of his time fighting in Long An Province. According to his military superiors and comrades, Richard was an exceptional soldier and leader. One of Richard's soldiers Paul Arca states, "I was Lieutenant Gwinn's RTO and spent about five months in the Bush with him. He was a great soldier and leader...I was proud to serve with him." Lieutenant Gwinn's unit was scheduled to return to Ft. Bragg, North Carolina by December 15, 1969; however, Lieutenant Gwinn did not return with his men. He was killed by hostile action fire on September 26.

Mack G. Gwinn recalls seeing Richard just before his last mission in Pine Apple, an area known for hostile combat action. Mack says that he warned Richard and his commanding officer that the Vietnamese in Pine Apple were infamous for wounding a platoon point man in hopes of luring out and killing military leadership.

His body was recovered and sent home to Quincy where his funeral arrangements were held in Quincy Arrangements by Adams Funeral Home. Accompanying his body was Lieutenant John Martindale. Surviving besides his parents was his brother Mack W. Gwinn, who had just returned from Vietnam to Ft. Bragg around the time of Lieutenant Gwinn's death.

Those who knew Richard best will never forget the warm hearted and compassionate young man that inspired so many to achieve excellence.

Richard Gwinn was married to Shela Hobson, NGC class of 1969. They had a daughter, Jenny.

Special thanks to Warren Kirbo, Tom McLaughlin, and Mack W. Gwinn Jr. for their contribution to this biography.

JOSEPH HILLMAN III
CLASS OF 1966

Name: Joseph Hillman III

Born: November 18, 1944

Hometown of Record: Piedmont, Alabama

High School: West Rome High School

NGC years: 1962 - 1966

Date of Death: July 22, 1969

He is honored on Panel 51W, Row 30 of the Vietnam Veterans Memorial.

Joseph Hillman III was born in Marietta, GA on November 11, 1944 to Mr. and Mrs. Joseph Hillman II. Joe grew up the oldest of three children in a family well versed in the military life. Joe's father entered the United States Army Air Corps in 1939, and in 1944 met his wife, Connie Bowen, on Anniston Army Airfield in Anniston, Alabama. Joe's father spent the next twenty-two years serving in the U.S.A.F., taking his family around the world. In 1950, Joe's brother, Vermer, was born, and in 1952 the youngest brother, Steve, came along.

Joe started his elementary education in Japan. Over the next eighteen years, the family would spend time in France, Alaska, and New Mexico. Joe grew up as a quiet, intelligent young boy who even early on displayed a knack for leadership. His mother recalls that he was an excellent student who read much. She remembers her husband taking Joe hunting, and while on the trip, Joe would be more interested in reading than the hunt.

Joe started his secondary education at Anchorage High School while living on Elmandorf Air Force Base. While in Anchorage, Joe's father participated on the pistol team at a local shooting range, an activity that sparked Joe's interest in sharp shooting. Joe's father recalls that Joe spent a lot of time on the range fine tuning his shooting skills, and in 1961 he achieved one of the most prestigious NRA competition honors. In 1961, the family relocated to Piedmont, Alabama where Joe graduated from West Rome High School in 1962.

Joe started North Georgia College in the fall quarter of 1963. While at North Georgia, Joe trained with Charlie Company in the Third Platoon, as well as Echo Co. in the Third Platoon. He participated on the Rifle Team, as well as the Officer's Club.

While at North Georgia, Joe became close with Ralph Colley. Ralph remembers Joe for his "menacing but not aggressive" nature. According to Ralph, Joe took on any challenge with a vigorous attitude, and he maintained emotional distance from most of his classmates. Joe took his military duties seriously, and looked forward immensely to serving his country in the United

States Army as an infantry platoon leader, a position that not many North Georgia graduates desired. Chad states, "Only about twenty percent of our class went infantry. If someone looked pissed off, as Joe often did, he went infantry." Joe graduated from North Georgia in the spring of 1967.

Tragic events in the fall of 1967 weighed heavy on Joe. In October, Ralph and his fiancé Betty Ann were scheduled to be married, and Joe was to be the best man. Joe was also to serve as a groomsman in another of his friend's wedding in the same year; however, before Ralph's wedding many members of the wedding party for the other wedding Joe was to attend were killed in a tragic car accident. Furthermore, Joe's own fiancé broke it off with him. These events left Joe emotionally bankrupt, and the pain kept Joe from attending Ralph's wedding. However, the two would hook up again at Fort Campbell, where the newly commissioned lieutenants were roommates before going to Vietnam.

Lieutenant Hillman began his tour in Vietnam on December 3, 1967 where he served as a rifle platoon leader in the Third Brigade of the 101st. Joe led his men by example, and his menacing attitude kept him on the frontlines of combat duty. He spent a lot of time fighting in Trang Bang in the Cu Chi area, a city made infamous for its vast and complex tunnel systems that aided the Vietnamese tremendously in causing significant U.S. casualties. While in Vietnam, Ralph and Joe hooked up for the last time on Christmas day during the culmination of the three Battalion Prop Blast Parties. Ralph recalls, "I remember little other than we couldn't stop giggling... all in all not a bad way to remember Joe, all a giggle with a big grin under those black rimmed glasses."

In July of 1967, Ralph commanded Charlie Company, and Joe led a Rifle Platoon on a mission back filling for the 25th Infantry around Cu Chi. On July 21, both Ralph's and Joe's companies participated in a six company air assault to place a cordon around a cluster of villages. The next day, on July 22, Lieutenant Hillman lost his life by a single AK47 round to the forehead.

The twenty-three year old First Lieutenant's body was recovered and sent back to Piedmont, Alabama where his funeral services were held at the Rock Run Baptist Church, with the Rev. J.R. Hawthorne and a military chaplain officiating. He is survived by his mother and father, and both of his brothers. His body rests in Oak Knoll Cemetery in Floyd County, Georgia.

For his exemplary service in Vietnam, Lieutenant Joe Hillman III earned the Silver Star, the Bronze Star, and two Purple Hearts. However, his lifetime achievement lies in the legacy of courage and leadership he inspired in those that served with him. The spirit of Joe Hillman is best captured by Ralph Colley, "The Joe we knew was a man of honor, a patriot by choice and conviction, fearless in conduct and passionate about his life's interest."

Special thanks to Mr. and Mrs. Joseph Hillman II and Ralph Colley for their contribution to this biography.

William H. Hunt
Class of 1968

Name: William H. Hunt

Born: November 6, 1946

Hometown of Record: Merritt Island, Florida

High School: Thomasville High School

NGC years: 1964 - 1968

Date of Death: February 25, 1969

He is honored on Panel 31W, Row 49 of the Vietnam Veterans Memorial.

Born on November 6, 1946 in New York State, William Howard Hunt, or Bill as he was known by those closest to him, grew up the oldest of three children. Bill's father, Richard Hunt, served in the United States Army before starting a career as the manager of an insurance company in Merritt Island, Florida, and his mother stayed at home to care for him as well as his brother and sister, Sandra and Larry.

Growing up, Bill was a neat and organized boy. He attended elementary school in Gainesville, Florida and spent some time at Gainesville High School before moving to Thomasville, GA where he graduated from Thomasville High School in 1964. While in high school, Bill began his lifelong love of weight training and body building. His sister still remembers the homemade bench that Bill used for lifting weights.

In 1964, Bill enrolled for the fall quarter at North Georgia College. In his freshman year, he trained with Golf Company in the First Platoon. In 1966, Bill transferred to Foxtrot Company where he trained with the First Platoon before moving to Charlie Company, Second Platoon in 1967. Although Bill enjoyed the cadet life with his classmates and Sigma Theta Fraternity brothers, he considered his military training duties his number one priority, and he was extremely dedicated to becoming the best soldier he could be. He spent much of his time at North Georgia in the weight room, training his body to a fine tuned athletic condition. It was important to Bill that he maintained a sound mind and body. Tony Faiia, a classmate and close friend of Bill, states, "He was an average student, serious minded, straightforward, and a hardworking guy."

In the spring of 1968, Bill and his closest classmates met in Camden, South Carolina for the marriage of Bill and his fiancé Nell Chance. After his marriage, Bill returned to Dahlonega a happy newlywed, and in 1968 he served as a Platoon Commander in Bravo Company before graduating in the spring with a B.S. in Business Administration.

Early in Bill's college life, he elected to join the Marine Corp rather than signing a contract with the Army. After graduating from

North Georgia, Bill entered the Marines as a Second Lieutenant, where he headed a platoon as a Basic Infantry Officer. Lieutenant Hunt began his tour in South Vietnam on December 21, 1968.

While in Vietnam, Lieutenant Hunt commanded a company on Firebase Russell, in the Khe Sanh area, approximately four miles from the demilitarized buffer zone. In late February 1969, members of the North Vietnamese Army slipped past the Marine guarding a fox hole at the base of the mountain upon which Firebase Russell sat. The NVA climbed the narrowest point of the mountain and began to lay siege on Lieutenant Hunt's base. On February 25, the twenty-two year old lieutenant lost his life when a satchel charge was thrown into his command bunker. Three days later, a company from neighboring Firebase Alpine was airlifted into Russell to relieve the surviving Marines. Of the approximately 120 Marines at Russell when the attack began, 90 either lost their life or were wounded.

Lieutenant Hunt's body was accompanied home by classmate and fellow Marine Jim Ruska. Once again the former North Georgia Cadets assembled, but this time in Sumter, South Carolina to see that their good friend and United States Marine Corp officer received a proper burial.

William Hunt is dead, but the memory of this fine young Marine will live on in the hearts and minds of the numerous people who were influenced by the passion and leadership he shared with them. "May God rest his soul and look out for his family." Tony Faiia, NGC '68

Special thanks to Jim Ruska, Tony Faiia, Ron Kelley, and Sandra Hunt for their contribution to this biography.

CHARLES BUFORD JOHNSON, JR.
CLASS OF 1957

Name: Charles Buford Johnson, Jr.

Born: April 17, 1938

Hometown of Record: Oglethorpe, Georgia

High School: Oglethorpe High School

NGC years: 1953 - 1957

Date of Death: June 14, 1963

He is honored on Panel 1E, Row 23 of the Vietnam Veterans Memorial.

Charles Buford Johnson was born on April 7, 1938, in Macon County, Georgia. His parents were Charles Buford Johnson, Sr., a worker at the Montezuma Post Office and an owner of a small dairy farm, and Florence Perry Johnson, a local schoolteacher. Charles spent his childhood in Oglethorpe and his first job was on the family farm. Charlie graduated from Oglethorpe High School and then went to North Georgia College.

At North Georgia College, Charlie's career really blossomed. He was involved in many student organizations such as the varsity basketball team, the Rex fraternity, the intramural football team, the officer's club, the YMCA, and the Letterman's Club. In the Corp of Cadets, he became a captain and the 1st Battalion Commander. Not only was Charlie an active student, he was also well liked.

Many of his classmates remember him fondly. His nickname on campus was Foggy. This was because he always seemed to be a very kind man and seemed to be lost in a fog. Many of the men under his command referred to him as a competent leader but never pushy - always there and willing to help. In 1957, he graduated from North Georgia College with a BA in History and went out to pursue his military career.

It was sometime after graduation that Charlie met his wife, Constance "Donnie" Johnson. The two of them met in Puerto Rico and had a daughter named Caroline Bernadette Johnson. Soon after, Johnson headed to Vietnam. About the same time he was leaving for Vietnam he would have a final encounter with an old classmate, Sanders Hale. At the time, Hale had been on active duty for six months at Fort Benning. As he was loading his bags, Johnson pulled up in his car, and the two talked for a while. As they parted ways, Johnson said he was heading for Vietnam. This would be one of the last times Charlie was seen alive by a former classmate.

In Vietnam, Charlie was one of the first Americans in the county. He was a military advisor in the MAAG, Military Assistance Advisory Group, there to advise the Vietnamese in the conflict. At the time he was killed, his group wasn't even armed. They were travelling through the country. On June 14, 1963, he was shot by a sniper while he was a passenger in a jeep, making him the first Georgian killed in the war.

Charlie's death affected many people back home; many of his classmates were saddened by the news, he was always known as a gentle guy, and it really shocked them to hear what had happened. Worse than that was how his family took it. Charlie was an only child, and his parents took the news hard. The town doctor, also a family friend, said, "That bullet that day didn't just kill Charlie, it killed Florence and Buford as well – three people died that day." However, despite the trauma caused by his death, his family lives on. He has a lot of family still living in Macon County, and his daughter is currently living in North Carolina.

Thanks to Elizabeth Mahoney, Sanders Hale, Haines Hill, and the North Georgia Alumni Association for their help in producing this biography.

Charles Johnson as a child with his mother
(Photograph provided by Elizabeth Mahoney)

MILO P. JOHNSON
CLASS OF 1961

Name: Milo P. Johnson

Born: October 2, 1939

Hometown: Augusta GA

High School: Unknown

NGCSU years: 1958 – 1961

Date of Death: September 1, 1967

He is honored on Panel 25E, Row 78 of the Vietnam Veterans Memorial.

Milo P. Johnson was born on October 2, 1939 in Augusta GA. He attended North Georgia College and graduated in the Class of 1961. He received a B.A. in Biology. While at North Georgia, he was a member of Who's Who, Biology Club, Rex, Drill Platoon, NCO Club, Officers Club, Scabbard and Blade Society, YMCA, DMS, and he made the Dean's list. He was also the commander of Alpha Company.

Billy York, one of his freshman friends, said of him, "Milo was the greatest Company Commander I ever had. He was a great leader and very well rounded. He had a way of making even the most insignificant freshman feel important, and he would do anything to protect his men. He was kind, sensitive, firm, and bright."

Billy York goes on to tell that while at North Georgia, Milo spent the second half of his senior year on room confinement because he took the company down to Lake Lanier and had a beer party. The whole thing was supervised, he said, but somehow the commandant staff found out and burned Milo. Milo would take the burn slips and post them on his door as a sense of accomplishment and pride.

This is how he was as a person - He always looked for the good in a person and a situation, and he wanted to have a good time. He was commissioned in the Army and joined the Special Forces, where he attained the rank of Captain. While in Vietnam, he worked with local mountain warriors. While trying to break up a fight, he was shot. Every person who new and met Milo said the same thing: he was a great and caring guy and a natural born leader. Milo is survived by his wife and daughter.

FRANCIS McDOWALL, JR.
ATTENDED 1963-1964

Name: Francis McDowall, Jr.

Date of Birth: September 3, 1944

Hometown of Record: Lawrenceville, GA

High School: Lawrenceville High School

NGC years: 1963 - 1964

Date of Death: August 12, 1969

He is honored on Panel 19W, Row 18 of the Vietnam Veterans Memorial.

Francis McDowall, Jr. was born on September 3, 1944. He grew up in Lawrenceville, Georgia. After graduating high school, he was compelled by a wish to serve his country and attended North Georgia College. He began attending North Georgia College in 1963. During his time at school, he served as a cadet in the school's Second Battle Group, and more specifically, he was in Foxtrot Company. He left college in 1964 to join the United States Army. He attended Aviation School at Ft. Rucker, Alabama and graduated in 1968 as a helicopter pilot.

After being trained to fly Cobra helicopters, he was deployed in November of 1968 to Camp Evans, which was located near the North Vietnam border. He served in the 1st Calvary Division, 2nd Battalion, 20th Field Artillery as a Warrant Officer, First Class.

Francis McDowall, called "Mac" by his friends and fellow soldiers, was trained for his new job and tasks by Robert "Bob" Hartley. When McDowall arrived in Vietnam, he served as protégé to Hartley in order to be trained to fly Cobra helicopters. McDowall was described as "very quick, smart" and he "always asked questions" about his new line of work. As he learned about his new helicopter, unlike his fellow pilots, he enjoyed working on the mechanical aspects of the helicopter. Hartley thought he fit in well in his unit and believed that he did an "excellent job". McDowall had an enormous appetite to learn as much as possible about a Cobra helicopter and whatever new mission he faced.

His unit's primary task was to provide aerial support to troops on the ground through the firepower of a Cobra helicopter. While ground troops' existence was characterized by hours of boredom punctuated by moments of sheer terror, the task of the Aerial Rocket Artillery was far more regular. The Cobra helicopter pilots would fly into combat to provide fire support for the airlift helicopters that flew troops in or evacuated them out.

Each battery of helicopters was subdivided into three platoons, each with four helicopters prepared for battle at a moment's notice. According to Hartley, their platoon was capable of getting a helicopter off the ground in forty-five seconds flat. For each mission, two helicopters would fly in unison. These helicopters could deliver a payload of firepower that the Viet Cong could not hope to match. Each helicopter carried seventy-eight rockets, ten to seventeen pounds apiece, and after their payload was loosed, the pilots would return to base to be reloaded and refueled. Both men loved their job. They were able to engage in combat regularly and from afar.

Their unit was moved to Qan Loi, north of Saigon near the Cambodian border. Their mission, however, remained the same – to provide strategic air support. In April of 1969, Hartley's service in Vietnam ended, and he returned to the States. Through his ability to learn and lead during this challenging time, McDowall earned a promotion to the position of Pilot and Aircraft Commander and directly led missions against the Viet Cong.

Another of McDowall's friends that served with him in Vietnam, Paul Yacovitch, arrived at Qan Loi in February of 1969 and was assigned to McDowall's unit. At the time of Yacovitch's arrival, McDowall was already considered a seasoned combat pilot who was an expert in his line of work. Yacovitch was trained by McDowall and served as his copilot and protégé. Eventually Yacovitch piloted his own Cobra, and the two men would often fly on missions together.

On their final mission together, McDowall and Yanovitch were scrambled by the Tactical Operation Center, and the Cobras were off the ground in no time. They were sent to another landing zone thirty miles away from Qan Loi to provide support to their home base should it be hit. Their commander was correct, and Qan Loi was nearly overrun by Viet Cong. The base was hit by a major coordinated attack. The two helicopters lifted off and returned to their base to supply support fire for their troops on the ground. The Viet Cong broke through the perimeter and were soon fighting in the base itself. The two Cobra pilots arrived just as the action escalated and loosed their payloads on their foes. The Cobras were able to avoid most enemy fire thanks to their ability to dive in from high altitudes, and the ability to rapidly deliver massive amounts of ordinance suppressed the Viet Cong effectively. Regrettably for McDowall and for other helicopter pilots in Vietnam, the Viet Cong had been armed with rocket-propelled grenades which countered the Cobra helicopter.

As dawn approached, McDowall's helicopter had run out of ammunition and was soon to be running out of fuel. He desperately needed to land his helicopter. The two copters hovered above the camp and at dawn on August 12, 1969, they saw an opportunity to land. As they prepared to park their helicopters, Qan Loi again came under attack. The two successfully landed their copters, and as McDowall prepared to debark, his helicopter was hit by a rocket-propelled grenade. He did not even know what hit him. His copilot was far more fortunate and was literally blown clear of the combat. Paul Yacovitch remembered this moment as if it was yesterday and would like us all to remember the sacrifice McDowall made for his country. Today there stands a memorial for McDowall's service at Hartsfield-Jackson International Airport in Atlanta.

For all the information provided, a special thank you goes out to Robert Hartley, Mike Sheuerman, Gary Roush, and Paul Yacovitch.

LARRON D. MURPHY
CLASS OF 1967

Name: Larron D. Murphy

Born: October 5, 1944

Hometown of Record: Dalton, Georgia

High School: Dalton High School

NGC years: 1962-1967

Date of Death: April 23, 1970

He is honored on Panel 11W, Row 41 of the Vietnam Veterans Memorial.

Larron D. Murphy was born on October 5, 1944 in Dalton, Georgia. He attended North Georgia College and was a part of the Sigma Theta fraternity. He left NGC before graduation and served in the Army as a Rotary Wing Aviation Unit Commander.

On April 23, 1970, his helicopter crashed on a search and rescue mission. There were two AH-1G and two UH-1H aircraft on the mission to fly into enemy territory and retrieve a long-range reconnaissance patrol. Half of the aircraft had to abort the mission. The UH-1H aircraft aborted the mission, but the other two aircraft continued on to provide covering fire to the patrol so that they could retreat to friendly lines. After a few minutes, Murphy's AH-1G and the other spotted the long range patrol. Thirty seconds later, Murphy's aircraft reported that it was crashing.

The next day a search party was dispatched towards the reported crash site. The crew of the other AH-1G aircraft located the crash site by memory, but the search party was never able to locate any remains of the crew or the AH-1G aircraft. Murphy was not the pilot of the AH-1G aircraft. The pilot, Dennis K. Eads, was also killed in the crash. The reason that a crash site was never found was that the helicopter exploded in mid-air during the rescue mission. Until 1975 he was considered missing in action, and after missing for five years, the Army pronounced Murphy dead. His body was never found.

Roy Lynwood Murphy
Attended 1962-1963

Name: Roy Lynwood Murphy

Born: February 13, 1944

Hometown of Record: Moultrie, Georgia

High School: Moultrie High School

NGCSU years: 1962 – 1963

Date of Death: May 29, 1965

He is honored on Panel 1E, Row 128 of the Vietnam Veterans Memorial.

Jesus said, "Greater love hath no man than this, that a man lay down his life for his friends." This statement, written some 2,000 years ago, provides a measure of understanding when examining the sacrifices made by the soldiers and Marines who fought the Vietnam War. One such Marine, Lance Corporal Roy Lynwood Murphy, made that sacrifice and became a hero to his squad mates. Roy could only make this sacrifice because of the life that brought him to the point of his death. His bravery in the Vietnam War represents the culmination of his life at home, his time at North Georgia College, and his time with the military.

Lance Corporal Roy Lynwood Murphy represented the idyllic small town kid that stories steeped in Americana idolized. He was the real life Opie Taylor. He was born February 13, 1944 in the small South Georgia town of Moultrie. He embraced the ideals of small town Georgia life. According to Nick Wiltse, Roy's best friend, he only dreamed of returning to the States to raise cattle and be near his family. Mr. Wiltse spoke about Roy winning a ribbon for cattle. At the time of his death, his parents, Mr. and Mrs. Roy C. Murphy; two brothers, Keith and Mark Murphy; and his grandparents, Mr. and Mrs. G.W. Mims survived him. These family members all lived in Moultrie.

Upon graduating high school, Roy enrolled in North Georgia College, the Military College of Georgia. His stint at North Georgia College only lasted one year. His freshman class began in 1962. As a member of the Corps of Cadets, he belonged to Alpha Company, second platoon. This information can be found in the 1963 *Cyclops* alongside his picture.

Lance Corporal Murphy left North Georgia after one year to pursue a military career. Roy enlisted with the Marine Corps. The Marine Corps fought both the guerilla warfare of the Viet Cong, and the more conventional warfare of NVA, or North Vietnamese Army. In preparation for this kind of fighting, Roy went to Okinawa, Japan.

Nick Wiltse, Roy's best friend and fellow Marine, attended the Amphibious Raid School with him. Nick said that they trained for a Raider Platoon in anticipation of going to Vietnam. One of his squad mates, Craig A. Slaughter, tells us that Roy served with "Mike" Company, 3rd Battalion, 3rd Regiment, 3rd Marine Division. 3rd Marine Division received the task of securing the area around the Chu Lai Airstrip. To accomplish this goal, the Marines performed several different exercises. Search-and-destroy missions consisted of Marines sweeping through villages to destroy weapon caches. The Marines also executed night ambushes. These missions involved Marine units patrolling the jungle to curb enemy activity. Lance-Corporal Murphy lost his life on a night ambush. Roy's friend, Nick Wiltse, provides the narrative:

Seven to nine Marines were on an ambush mission. We were patrolling after dark. In the early morning hours, grenade and small arms fire erupted from the jungle. Roy stood up and exposed himself to the gunfire and returned fire on the Viet Cong, hoping to protect his squad mates.

He died on May 29, 1965, at the young age of twenty-one. His actions that fateful night are indicative of the kind words spoken about him. Master Sergeant Benjamin Sandoval shares the account this way, "Roy was my team leader and I was with him the night we were ambushed. I will never forget that night because he had told me that he was taking the tail end of the patrol which was normally the position which I assumed. I feel that I should have been in his place and not him." His actions afforded him the honors of a Purple Heart, the Combat Action Ribbon, the Vietnam Campaign Medal, the Presidential Unit Citation, the Good Conduct Medal, and the Armed Forces Expeditionary Medal.

Though his life ended too quickly, Roy exemplified the idyllic Marine. He quickly responded to the threat and provided an opportunity for his fellow Marines. As that jungle erupted in a hail of bullets and grenades, Roy knew one thing; he knew that his fellow Marines needed him. The Marine slogan, Semper Fi, means "always faithful." For Marines, this slogan applies to both the

Corps and Country. In May 1965, Roy showed his understanding of "Semper Fi."

Raid School Okinawa 1965: Contributed by Nick Wiltse

Roy, Fisher and Pugh: Contributed by Nick Wiltse

Roy's Signed Picture: Contributed by Nick Wiltse

JOHN RUDOLPH PEARSON
CLASS OF 1956

Name: John Rudolph Pearson

Born: April 27, 1934

Hometown of Record: Thomasville, GA

High School:

NGC years: 1952-1956

Date of Death: August 17, 1966

He is honored on Panel 10E, Row 16 of the Vietnam Veterans Memorial.

John Rudolph Pearson graduated from North Georgia with the Class of 1956 where he was a big hit amongst students and faculty. The following is an excerpt from our very own *Cyclops*:

"Peaches" Pearson is one of the most popular boys on the campus, besides being one of the most dependable. He was elected to "Who's Who in American Colleges and Universities" because of his well-rounded personality. He was active in Sigma Theta as Sergeant-at-Arms, the NCO Club, Officer's Club, B. A. Club, and as Executive Officer in Scabbard and Blade. He was also President of the Sophomore Class, a member of the Pan- Hellenic Council, and Band Company Commander. He will receive his B.S. Degree in Business Administration.

Although his branching results are scarce, he did end up entering into the Special Forces branch. While overseas, he attained the rank of Major and was moved into the Detachment Executive Officer position in Detachment B-25 of the 5th Special Forces Group, which operated around Duc Co in Pleiku Province, South Vietnam at the time of his promotion. This Special Forces camp had been the site of a major assault by North Vietnamese and Vietcong forces some months before and continued to be a hot spot for enemy contact. While the exact details of MAJ Pearson's death remain classified, the KIA report lists him as suffering from "Multi-fragment wounds" on 17 August 1966. This would signify either a grenade, mine, mortar, or artillery shell as the most probable cause of his death.

Not many soldiers can answer the call of duty that is implied by attempting to join the Special Forces. MAJ Pearson did answer the call, however, and did his duty no matter the cost and paid the ultimate sacrifice for it. MAJ Pearson will go down in the scrolls of history as a hero to the American nation and a mark of pride to North Georgia College and State University.

ROBERT L. PHILLIPS
CLASS OF 1968

Name: Robert L. Phillips

Born: August 10, 1946

Hometown of Record: Oxford, Georgia

High School: Oxford High School

NGC years: 1964-1968

Date of Death: May 6, 1970

He is honored on Panel 11W, Row 116 of the Vietnam Veterans Memorial.

Robert Littleton Phillips was born on August 10, 1945, in Oxford, Georgia. He attended North Georgia College and graduated in 1968. While attending North Georgia College, he accrued several accolades. Murphy was named Mr. NGS during his senior year. He was a part of the Corps of Cadets, he was part of the Scabbard and Blade, participated in Sigma Theta, and he was a member of the Glee Club.

While at North Georgia College, he learned the skills that he would need to become a lieutenant in the army. His company was fighting heavy resistance on the day of his death. He was killed in action on May 6, 1970.

Phillips' unit, along with several other units, was flying in to a landing zone to return to combat after a three day reprieve. Everything was cleared for the mission the night before by reconnaissance teams. As the mission started, it was soon discovered that the landing zone was overrun with North Vietnamese forces. The landing of additional helicopters into the landing zone was called off until the units that were there could re-establish superior firepower and control the zone.

Phillips was still in the air when the order to retreat was given. Somehow he convinced the pilot to land, and he said that there were good men down there and he wasn't going to stop until they were able to reach safety. After he convinced the pilot to land, he and his men in the chopper somehow managed to re-establish fire superiority over the North Vietnamese. He was laying smoke to signal the medical evacuation helicopters so that the wounded men from the ambush could be brought back to safety.

According to eye witnesses, Phillips was shot from behind after he walked past a Vietnamese spider hole door where soldiers were laying in wait for the American armed forces. He died that day to save his men.

His unit remembers him as a valiant leader that cared first for his men. This was a man that could have conquered the world but chose to serve his country. Phillips and his friends would continuously talk about how their greatest asset was their men and their men came before all else. He also talked about how North Georgia College had prepared them to do what was necessary to be successful in the military, in battle, and in life.

Robert Ira Rabb
Class of 1968

Name: Robert Ira Rabb

Born: October 16, 1946

Hometown of Record: Darien, GA

High School:

NGC years: 1964-1968

Date of Death: May 5, 1970

He is honored on Panel 11W, Row 105 of the Vietnam Veterans Memorial.

Robert Ira Rabb graduated from NGC and commissioned into the Infantry branch in 1964. At that time, aviators were pulled from their branches and would be supporting. He went through this program and was a graduate of flight school class 69-18. He was then assigned to HHC, 222nd Combat Aviation Battalion, 12th Combat Aviation Group, 1st Aviation Brigade and began his tour of Vietnam on August 15, 1969.

On May 5, 1970, recently having been assigned to a gun platoon in the 195th Assault Helicopter Company, Rabb and his crew, consisting of CW2 Charles G. Dougan and SP4 James D. Smith, were conducting a secret mission across the border of Cambodia. Their tail number for their UH-1C was 64-14120. Army records have them operating west of Quan Loi at the time of their mission. During their mission they were hit with a rocket propelled grenade and the helicopter exploded three times, according to eyewitness accounts. Because they were conducting an operation that was secret at the time, the crash was reported as being within Vietnamese borders. The area was considered very hostile, so initially, no attempt was organized to immediately go recover the bodies. They were, however, recovered later. 1LT Rabb was identified by his MACV card.

Rabb's great sacrifice to this nation and for the freedom of people in Vietnam will not be forgotten, especially here at North Georgia College.

ROBERT NED SAULS
ATTENDED 1959-1960

Name: Robert Ned Sauls

Born: September 12, 1941

Hometown of Record: East Point, Georgia

High School: Sylvan Hills High School

NGC years: 1959-1960

Date of Death: March 11, 1969

He is honored on Panel 29W, Row 14 of the Vietnam Veterans Memorial.

Joseph Campbell once said, "A hero is someone who has given his or her life to something bigger than oneself." Robert Ned Sauls knew what it meant to be a hero. Robert held strong convictions about what it meant to be an American, and part of those convictions included his views on military service. These beliefs led him to North Georgia College and then to the Army. He served the American public for a very long time in the army before he was killed in Vietnam.

According to the *Cyclops*, Robert came from East Point, Georgia. Alice Keith Collier, the woman that he was engaged to for eight months, said that he had been adopted by his stepfather. His original name was Robert Ned Williams. Loyse Albert Sauls and Mary Ruth Sauls were his parents. L.A. "Rip" Sauls was a military man for over twenty years. Perhaps this fact played a role in Robert's decision to join the military. He graduated from Sylvan Hills High School in 1959. John Adams, a high school classmate says, "We remember your dedication to your ROTC classes and knew you would make an excellent soldier." Some time after he and Alice ended their engagement, he married a woman named Nancy. Robert and Nancy Sauls had a son, also named Robert.

Robert Ned Sauls entered North Georgia College in 1959 right after high school graduation. He only stayed at the school for one year. While at the school he became involved with the Blue Ridge Rifles, North Georgia's elite rifle team. He was assigned to Charlie Company. His picture appears both with the Rifle Team and with Charlie Company's photographs in the 1960 *Cyclops*.

At the conclusion of his freshman year, Robert Ned Sauls enlisted with the Army. He actually served for eight years before he died in 1969. He attained the rank of Chief Warrant Officer and was a target acquisition radar technician. He served with the Headquarters and Headquarters Battery, 8th Battalion, 26th Artillery, 254th Field Artillery Detachment, I Field Force.

He died March 11, 1969. The Distinguished Service Cross Citation notes his bravery. To paraphrase the citation, CWO Sauls exposed himself to the enemy barrage during a siege, located many enemy emplacements, braved fire to evacuate casualties, and directed gunships against communist forces. His heroics continued when he drove his jeep through the battle to restock ammunition, received fragment wounds from a mortar, and repeatedly entered a burning bunker to save lives and supplies. Finally on March 10, 1969, he raced to the scene of a devastating skirmish and began helping all the wounded. It was this last heroic effort that cost him his life.

While he tried to save an injured Vietnamese officer, a hostile round struck nearby and fatally wounded him. He died the next day. For his services rendered to his country, Chief Warrant Officer Robert Ned Sauls earned the Distinguished Service Cross, Purple Heart, National Defense Medal, Vietnam Service Medal, and the Vietnam Campaign Medal.

Robert Sauls believed in his country and sought to do his best to serve the country that he loved. During their engagement, he sent the "American's Creed" to Alice Keith. This statement boldly represented what he believed to be his duty. The Creed says, "I therefore believe it is my duty to my Country to love it, to support its Constitution; to obey its laws; to respect its flag; and to defend it against all enemies."

Reading the Distinguished Service Cross Citation and thinking about this last statement, Robert Sauls loved America. He gave his life for the ideals that define our country. That citation reads like a movie script; his heroics over that two day stretch rival that of Hollywood's greatest action heroes. Yet, Robert was more than that, he embodied heroism. His script flowed from reality and not the mind of a screenwriter.

LEONARD HOWARD SMITH
ATTENDED
1963-1965

Name: Leonard Howard Smith

Born: Jul 26, 1945

Hometown of Record: La Grange, Georgia

High School:

NGC years: 1963 - 1965

Date of Death: April 25, 1967

He is honored on Panel 18E, Row 88 of the Vietnam Veterans Memorial.

Leonard Howard Smith and I were classmates from Mrs. Welch's kindergarten until our sophomore year at NGC. Howard, as we knew him, could have dinner with George Bush and supper with Fred (the homeless Vietnam Veteran who roams the downtown LaGrange area). Fred and George would be treated with the same respect. Howard had no inflated ego and would puncture yours.

Following is an article that appeared in *The LaGrange Daily News* on Veteran's Day approximately ten years ago. This article was submitted by Richard L. Sheridan, PhD along with a request that it be published on Veteran's Day "as a tribute to all veterans and especially my friend, Leonard Smith, who was from the LaGrange area." Sheridan writes:

We were both U.S. Marines serving together in Vietnam. Smitty was killed in combat in 1967, but I will never forget him. I want his family to know that Smitty lives on in the hearts of those who served with him in the 11th Engineer Battalion.

The Daily News was proud to honor Sheridan's request.

It was August, 1965. Our landing ship (LSD), bobbing in the South China Sea off the coast of Chu Lai, was stuffed with heavy equipment and several platoons of the 1st Engineer Battalion as part of the 7th Marine Regimental Landing Team. I really didn't know what we were doing there, nor do I to this day. Operation Starlight, it was called. The first major offensive movement of the Vietnam War. Daylight arrived and off-loading equipment down the ship's ramp began against a backdrop of F4 Phantoms streaking across the sky. I was scared as hell, but none of my fellow Marines would admit it to each other. I wondered if I would make it to my 19th birthday. I didn't want anything to happen to me because I knew my mother would never forgive herself for consenting to let me join the Marines a year earlier. We were heading into the monsoon season for a very long year, in a war that nobody wanted, against an enemy I did not know, in a land I had never heard of, and for reason that were less than clear. The luster of my new tattoo was beginning to wear off.

Late 1966 now, part of the 11th Engineer Battalion near Ca Lu and Dong Ha. The mission, as we understood it, was to widen the Demilitarized Zone, an imaginary strip of land separating North and South Vietnam. My friend, "Smitty," talked about his plans to go to college in his home state of Georgia after his year in the war was over, wanting to eventually become a dentist. My memory of him stopping a bullet while scrambling into his truck to get his flak jacket during an ambush still haunts me. I found his name, Leonard Howard Smith (Panel 18E, Line 88), and others like him among the 58,000 or so names engraved on The Wall in Washington, D.C.

The mission of this tragic war was never discussed among us, nor did we pretend to understand. After all, we were teenagers. We were in service to our country, and that was all that mattered. But, we were also feeling the anti-Vietnam War sentiment back home, thousands of miles away. We were being called names like "baby killers" and portrayed as a bunch of trigger-happy potheads. This was really puzzling since I didn't know one fellow Marine who used drugs my entire time in country. To be sure, innocent civilians were killed and there are things every combat veteran wished he had not done. But this was war, and war is not a sanitary process.

But what was the American public to expect? We were not a bunch of accountants choosing teams before going out for a weekend paintball gun frolic. This was not fantasy; it was war, and it was for real. The team included the mean fellow from the Bronx avoiding a court sentence, the kid from Iowa itching to see the world, the black guy from urban L.A. escaping poverty, the Latino from San Antonio proving his patriotism, and the Apache kid from Tucson dutifully fighting his country's wars like his ancestors have proudly done for the past century. No one would have confused us with a cultural diversity or a love fest, just Americans proudly serving the country we held in common. No student deferments needed in this bunch. Few had the money or inclination to go to college anyway. Our country was at the height of the Cold War, and fighting the spread of communism in Vietnam was a plausible rationale for being there. Well, at least for us it was.

It was the summer of 1967 now, and I added myself to the ranks of the increasingly disenfranchised war veterans of this country. There was no homecoming, no parade, no mention of a job well done. There were only looks of condemnation and disdain from fellow students active in the anti-war movement. I sensed that my fellow college students not only hated the war but also the warrior. It occurred to me that lepers and child molesters were better accepted in our country than we were. I tried to defend the importance of the U.S. involvement in Vietnam to my classmates. You know - the old yarn about stopping the spread of communism throughout the free world. This was a hard sell, and eventually, even I no longer believed that argument. It was easier for the Vietnam veteran on campus to conceal his involvement in the war than to tolerate contempt from others. I even had my "USMC" tattoo removed, further distancing myself from the reality of having fought in a war that most everyone hated.

This was the beginning of the period that Vietnam veterans refer to as "The Missing Ten Years." It was during this time that Vietnam veterans were alienated from their homeland, feeling that the sacrifices they made were in vain, and perhaps they were. It was the long silent message from the U.S. public to Vietnam veterans conveying scorn, shame, and indignation.

At least it seems different now. Time has a way of healing our country's war wounds. Vietnam veterans feel they are almost home, never held in adulation, but a least experiencing the illusion of being home. But Vietnam Veterans never wanted a parade anyway.

All we wanted was a scintilla of dignity and the right to self-respect. I would like to think that the most important lesson gleaned from the Vietnam War is how not to treat our country's men and women who serve in time of war, no matter how controversial or unpopular the war may be. Make no mistake about it; Vietnam was an ill-fated war fought for specious reasons. Nevertheless, the Veterans deserved a hell of a lot better than the shabby treatment they received during and after the war. Vietnam veterans, and all veterans, wanted peace like everyone else, but never will we forsake those who served in war for this country, particularly those who made the ultimate sacrifice.

Leonard would have been a damn good dentist.

BENNY THOMAS STOWERS
ATTENDED 1951-1952

Name: Benny Thomas Stowers

Born: March 4, 1934

Hometown of Record: Dawsonville, GA

High School: Dawsonville High School

NGC years: 1951-1952

Date of Death: May 18, 1966

He is honored on Panel 8E, Row 9 of the Vietnam Veterans Memorial.

Benny Stowers was born in Juno, Georgia, a small community in Dawson County, on March 4, 1934 and was the second of thirteen children. He spent his childhood in the area and attended a one room school house in the Dougherty area. Later, he attended Dawsonville High School and graduated in 1950. When Benny graduated high school he had two goals: to fly airplanes and to farm. This is also where Benny would meet his future wife, Hazel Burt.

After Benny graduated high school, he started to work for his uncle as a stone mason. The work was hard, and in 1950 he decided to go to school in order to find an easier way to make a living. In 1951 he applied to North Georgia, making him the first in his family to attend college.

He started at North Georgia in 1951 but was only there for a year. After North Georgia, he transferred to the University of Georgia. At UGA, he was one of the first people in the poultry science program. He also entered the ROTC program. At UGA in 1953, Benny married his high school sweetheart, Hazel. In 1955 he graduated from UGA with a degree in poultry science and gained his commission in the Air Force. 1955 was also the year that his daughter, Pam, was born.

In 1956, Benny went to Selma, Alabama where he started flight school and then moved to Texas where he started flying. While in Texas Benny's second child, Reggie, was born. After Benny completed flight school, his family moved to Albany, GA, and he went to Korea. Benny returned from his tour of Korea in 1961 as a captain.

After Benny returned from Korea, his family moved to Merced, California. After staying there for a few years, the family moved back to Albany. In March of 1966, Benny was sent to Okinawa where he was to fly the KC-135 midair refueling jets to refuel the B-52s on their bombing runs over North Vietnam. On May 18, 1966, two months into his tour, he was killed in a plane crash. The plane took off during a storm and crashed near the runway. He was buried at Juno Baptist Church in Dawson County, GA.

After Benny's death, his family moved back to Dawson County, where they live today. Currently, three of Benny's four grandchildren attend North Georgia College and State University.

I would like to thank Reggie Stowers for all of the help he has given me on this project.

ROBERT "BO" ACQUINN THOMPSON
CLASS OF 1961

Name: Robert "Bo" Acquinn Thompson

Born: July 11, 1939

Hometown of Record: Lincolnton, Georgia

High School: Avondale High School

NGC years: 1957 - 1961

Date of Death: August 9, 1967

He is honored on Panel 24E, Row 97 of the Vietnam Veterans Memorial.

Robert A. Thompson was born on July 11, 1939. To his friends, he was known as "Bo." He grew up in Lincolnton, Georgia. He attended high school at Avondale High School, where he graduated in 1957. After finishing high school, he joined the corps of cadets at North Georgia College and began work on a Bachelor of Arts in History. During his time in college, he met his wife to be, Hilda Hammond. They made such a great couple that, for a senior superlative, they were voted the "cutest couple" at North Georgia for the graduating class of 1962.

Hilda also introduced Bo to Billy York, who was to graduate in 1964. Hilda and Billy grew up together in Lincolnton.

Originally, Thompson would have graduated in 1961, but he came down with mononucleosis which forced him to drop out for his last two quarters of his junior year. During his time at North Georgia College, he enjoyed playing intramural football and softball and served on the color guard with his close friend "Banjo" Davis. Thompson and Davis became best friends thanks to their time spent together in Delta Company.

Thompson also joined the Sigma Theta Fraternity and participated in the Officer's Club. Davis and Thompson also enjoyed experiencing Summer Camp together. By the end of his career at North Georgia, he had reached the rank of Assistant S-3 in the Second Battle Group through hard work and dedication. He finished school after the fall of 1961. Hilda graduated in the spring of 1962, and they were married in November of that year.

Just before Bo and Hilda were to be married, he was stationed in the Florida Keys because of the Cuban Missile Crisis. In the advent of war, his division was suited up to fly into Cuba. Fortunately, the situation never escalated to armed conflict. After the crisis, Bo Thompson, Banjo Davis, and their wives moved to Ft. Rucker, Alabama where they attended flight school. The two men parted ways. Davis went on to serve in Korea, while Thompson was to serve in Vietnam.

Thompson joined the 145th Aviation Battalion, 120th Aviation Company. Thompson arrived in Saigon in September of

1963. Around the same time, John Givhan arrived in Vietnam. Thompson and Givhan made quick friends. Gihvan described him as being "a great guy with a wonderful sense of humor." For instance, Gihvan thought he outsmarted the competition when he purchased a rather fancy fifteen dollar suit in Saigon and wore it back to base. Wearing his new suit, Gihvan went into the Officer's Club to meet Thompson. Shortly after showing off his new suit, he bent over to pick something up and his pants ripped down the back. Apparently, "Bo laughed so hard that he nearly had to be hospitalized."

The two young soldiers also enjoyed betting on college football. In the fall of 1963, they bet on three games. Being an Auburn graduate, Gihvan naturally bet on his team. Unfortunately for Thompson, he lost all three of his bets and still remains several thousands of dollars in debt to Gihvan. Memories such as these remain some of Gihvan's favorite stories to this day, who still feels the loss of his "beloved copilot and the greatest man he ever knew."

The two bonded further during an intense moment in combat on April 12, 1964 at Flat Rock, Vietnam. Flying a UH-1B Huey gunship, Gihvan sat in the right seat, and Bo sat in the left. They were flying in tandem to drop their troops off, so that if one should lose control of the Huey, the other could easily pick it up. They appeared to escape combat, but were hit by a rocket as they approached 4,000 feet. The rocket exploded, badly damaging the aircraft, but did not disable it. Ghivan, however, was badly hit. His leg was severely damaged from the right kneecap down. Thompson was momentarily blinded when his visor, which he typically left open, slipped down and covered his eyes.

Despite his injury, Ghivan reached over and flipped Thompson's visor open. Thompson was then able to regain control of the helicopter. Ghivan was bleeding badly, but Thompson stopped the bleeding by holding the artery, which halted the blood flow from Ghivan's leg.

Thompson was injured as well, from shrapnel. Even with the circumstances as they were, Thompson was able to successfully land the helicopter one-handed. Med Evac picked them up. Essentially, they had saved each other's lives. The two arrived at the hospital in Saigon with Thompson covered in Ghivan's blood. At the hospital, a general was awarding Purple Hearts to those who were injured. Both were both personally given this award.

After his tour ended in the summer of 1964, Thompson returned home to Ft. Rucker where his first child, a baby girl named Kathy, was born on the 24th of September. His son John, who was named after John Gihvan, was born on March 4, 1965. Thompson returned to Vietnam in 1966 to serve on his second tour. This time he joined C Troop of 1st Squadron, 9th Cavalry Regiment of the 1st Cavalry Division. His job as a helicopter pilot remained the same, to provide airborne artillery support to ground missions.

In what would be his final mission, on August 9, 1967, he was to provide support to the movement of troops on the ground in the Song Re Valley, Quang Ngai Province, Vietnam. Ed Scholes listened into the conflict on the radio and described the fighting as "hot and heavy." The Second Battalion was hoping to gain control of a landing zone so that the army could control the surrounding countryside.

Troop C consisted of two Huey helicopters, which were piloted by Captain Thompson and by Major William Harvey. Harvey flew in front, in command of the mission, while Thompson flew in the chase position in the rear. The two were to provide aerial surveillance for the troops on the ground as well as ordinance support fire. At 9:45 a.m. Harvey's helicopter was flying at one hundred feet and was hit numerous times by fifty-caliber fire. He looked back to see Thompson's aircraft get hit so badly that it was set aflame and was sent perilously towards the ground. When the helicopter hit the ground, it exploded on impact. Harvey's helicopter also crashed, but his troops were able to escape with minor scratches and bruises.

Thompson and his crew were not so fortunate. There were no survivors. After Thompson's death, his body was recovered, and

Banjo Davis returned to Vietnam from Korea to escort it back to the national cemetery in Marietta, Georgia where he is buried. His burial was a patriotic one, with full regalia. His loss is still deeply felt by his widow Hilda Withers, Banjo Davis, John Thompson, and John Gihvan. Davis's son, Robert Thompson Davis, shares his name.

Today, his grandson, Brice Acquinn Thompson hopes to carry on the family legacy by attending West Point. Hilda Withers described her former husband as a "dedicated soldier who believed in what he was doing." A flame remains lit in front of the Lincolnton County Courthouse in hopes that his sacrifice will be eternally remembered.

A special thank you goes out to Ed Scholes, Gerald Lord, John Thompson, Myron "Banjo" Davis, Mike Sheuerman, Gary Roush, Billy York, Hilda Withers, and John Gihvan for all their contributions to Robert Thompson's tribute.

Robert A. Thompson is on the left. (Photo courtesy of Ed Scholes)

Robert P. Tidwell
Class of 1967

Name: Robert P. Tidwell

Born: November 10, 1945

Hometown of Record: Douglasville, Georgia

High School: Douglasville High School

NGC years: 1963 - 1967

Date of Death: April 29, 1969

He is honored on Panel 26W, Row 80 of the Vietnam Veterans Memorial.

Robert Paul Tidwell was born on November 10, 1945, in Douglassville, Georgia. He attended Douglassville High School. After graduating from Douglassville High School, he attended North Georgia College. He was said to be a quiet person. During his senior year at North Georgia, he married his wife, Emily. He graduated from North Georgia College in 1967, was commissioned into the Army, and after training, went to Vietnam. He was a First Lieutenant when he was killed during an ambush as he led his platoon within the Vietnamese jungle. It was said that he was a good leader and had the respect and loyalty of his men.

CHARLES ROSS WILLIAMS
CLASS OF 1958

Name: Charles Ross Williams

Born: August 13, 1936

Hometown of Record: Forsyth, Georgia

NGC years: 1954 – 1958

Date of Death: July 12, 1966

He is honored on Panel 9E, Row 18 of the Vietnam Veterans Memorial.

Charles Ross Williams graduated and was commissioned from NGCSU in 1958, after which he took part in the Army's flight school program. At that time, aviators were pulled from their branches and would be supporting. After completion of flight school, he was part of a joint effort between the Army and the Navy to seek out and destroy Vietcong or NVA forces and supply trains that used the river system of the Mekong Delta.

River Patrol Boats provided by the Navy and Seawolf Helicopters provided by the Army would launch from offshore naval vessels into the Delta on search and destroy missions.

Williams was taking off with an unidentified crewmember to perform one of these missions on July 12, 1966. It was a night take off from a naval barge. While lifting off, it was too dark for Williams to be able to locate the horizon. Because of this, he was not able to establish a rate of climb. About 10 seconds after his take off, Williams crashed into the water not far from the ship. His body was recovered soon after.

William's bravery in undertaking these missions is a strong mark of his character. His memory will ring true to the values instilled in him by North Georgia College.

David Beavers Wood
Class of 1969

Name: David Beavers Wood

Born: February 16, 1947

Hometown of Record: Douglasville, Georgia

High School: Douglas County High School

NGC years: 1965 - 1969

Date of Death: April 26, 1971

He is honored on Panel 3W, Row 14 of the Vietnam Veterans Memorial.

David Beavers Wood was born on February 16, 1947 in Douglasville, GA. David was the youngest child of Quinton and Betty Wood. His siblings, Amy and Frank, still reside in the state of Georgia. His sister, Amy, remembers that one of David's greatest accomplishments as a child was eleven years of perfect attendance at Sunday School, for which he received a pin. Amy recalls David as a very religious person who was deeply involved in Church.

John Hutcheson, NGC class of 1969, knew David from the 2nd grade until the end of David's life and remembers that one Sunday, in order not to miss Sunday School, he was brought in on a stretcher and up the fire escape into the church classroom. Hutcheson also remembers that the kids in school referred to David as "The General" because of all his pins from perfect attendance at Sunday School. He calls David an "interesting guy" and "the sweetest guy" and they often had classes together in grade school.

David's family knew that he was interested in the military at a very early age because some of his favorite toys as a child were small army figures. Because of this interest, North Georgia College was a choice that came to David easily. David loved history very much, specifically American History, which was his major in school.

After graduating in 1965 from Douglas County High School, David attended North Georgia College from 1965-1969. During his freshman year, he was a member of Foxtrot Company. The company received the Best Drilled Squad of the year award and placed 3rd in company progress. He transferred to Golf Company his sophomore year and stayed there for his final years at NGC.

David was active in organizations and was greatly appreciated by his schoolmates. As a member of the Spanish Club, the Aggressor Platoon, and first Treasurer and then later President of the Wesley Foundation, he illustrated many characteristics that displayed his leadership skills. He was also president of the recently formed Political Science Club.

While at school, Wood earned the nickname of "Snake" among his friends. The nickname arose from his uncanny ability to imitate the hissing of a snake. In the March 7, 1969 edition of *The Cadet Bugler*, Wood was one of the few students who was asked the question, "If you could change NGC, how would you do it?" Wood is quoted as responding, "Institute a merit system for advancement and reduction in the Cadet Corps, such as a qualification for rank."

While at North Georgia, Daniel E. Williams, Jr. was a great friend of David. They were roommates beginning when David was a junior. Williams describes him as "an easy-going guy with a good sense of humor. He never seemed interested in attaining 'rank' at NGC, but was very serious about the military and his obligation to serve. I remember him as having a calm and cool attitude, taking things in stride and maintaining a positive outlook." Williams also writes that, "David was an honest, sincere, dependable, and trustworthy person that I was proud to call my friend."

The alumni at the NGC class of 1969 reunion remember David fondly and eagerly shared memories of him. They described him as fun, level-headed, and mild-mannered.

David was commissioned and completed Armor Officers School at Fort Knox and was assigned to the 197th Infantry Brigade at Ft. Benning, GA. David also spent time at the Panama Canal before he left for Vietnam. While at Ft. Benning and serving in Vietnam, Wood sent Williams several letters and they corresponded back and forth. In a letter dated 22 April 1970, he said "Although I have little to no time of my own, the fulfillment of being a platoon leader, even in a leg unit, is about the best thing that has happened to me…"

While in Vietnam, David kept an extremely accurate journal and records. In the letters David sent home, he wrote about all the men that he looked after, including where they where from, who their families were, etc. It was very apparent that David took a special interest in the men in his unit. Often on the back of these letters David would draw a snake, a symbol of himself and his time at North Georgia College.

David applied for and was accepted to receive an early out (early leave from term of service in Vietnam). Instead of leaving in

October 1971, he would instead leave in August of that year. He applied for an early out because Valdosta State University accepted him into graduate school. His sister, Amy, believes his future goal was to pursue the career of a military chaplain.

David served with the F Troop, 17th Calvary, 196th Light Infantry Brigade, Americal Division, as a first lieutenant Platoon leader. In April 1971, the Brigade was relocated to Da Nang for security duties. On April 26, 1971, David's unit was on patrol in the Quang Tin Province with some ROK troops. After some time, David called for the patrol to halt and left to do some scouting. After a while, the medic that was with them, Paul Ferguson, saw black smoke and rushed to find David. After finding a man to cover him, they found David's body. The cause of death was determined to be a piece of artillery, rocket, or mortar that hit him.

Paul Ferguson remembers Wood as "a very popular leader and the men in his platoon respected him. We lost a great Platoon leader and man that day... He did not ask anyone to do something that he, himself, would not do." Nine days before his death, David wrote in his journal, "What ever beings control life are to be thanked for this entry in my journal. Surely my God is in this valley with me."

David's passing had a great effect on the whole family, Mrs. Wood especially. David's sister remarks that David's death changed her mother's personality forever, and she was never quite the same afterwards. On November 5, 1971, North Georgia College created the David B. Wood Memorial Scholarship Fund in accordance with the personal request of David's mother. It was this scholarship and the kindness of friends of David and friends of the family that helped somewhat ease the pain of the loss of David. For many years after, Mrs. Wood received cards and notes from David's friends.

On April 17, 2004, John Hutcheson spoke at a memorial service held at North Georgia. In his speech he said,

There were three boys from Douglas County High School who came to North Georgia in 1965: Robert Tidwell, David Wood, and John Hutcheson. Robert and David were first cousins. I knew them from the 2nd grade through graduation and commissioning at North Georgia. Their names are on this memorial, and I am standing here before you.

He went on to explain that while he "gave some" through his military efforts, David and Robert "gave all." In August of 2009, Hutcheson went on to say that, "He had a clean heart more so than anyone I have ever known".

David Wood was an exceptional person. He is continually honored in his hometown of Douglasville, GA through the Eternal Flame monument. This monument is dedicated to all veterans of Douglas County, Georgia who made the ultimate sacrifice through their service. Wood's name is one of the eleven listed under Vietnam.

Special thank you to Daniel Williams, Amy Huckaby, John Hutcheson, David Griscom, Leigh Blood, Jeanne Blood, Warren Kirbo, Mary Haigler, Bill Hackett, Paul Ferguson, and Terry W. Green for providing information in the research on this honoree.

David graduating at NGC (Photo by Amy Huckaby)

David Wood with his collection of names tags including a "Snoopy" nametag. (Photo by Daniel Williams)

About North Georgia College & State University

Motto: Truth and Wisdom
Established: 1873
Type: Public
President: David Potter
Students: 5,500
Undergraduates: 4,700
Postgraduates: 750
Location: Dahlonega, Georgia, United States
Campus: Rural; 112 acres (Main Campus); 722 acres (Total)
Athletics: NCAA Division II
Colors: Blue and White
Nickname: North Georgia, The Military College of Georgia
Mascot: Saint Bernard
Website: www.northgeorgia.edu

North Georgia College & State University, founded in 1873 and located in Dahlonega, GA, is the state's second-oldest public institution of higher education. Today, North Georgia serves more than 5,600 undergraduate and graduate students, including about 700 men and women in its Corps of Cadets.

One of only six senior military colleges in the United States, NGCSU is one of only six senior military colleges in the country in accordance with federal law - 10 US Code 2111a. The University System of Georgia Board of Regents has designated NGCSU as a state leadership institution for civilian and military students. Other senior military colleges include the Citadel, Norwich University, Texas A&M University, Virginia Military Institute, and Virginia Tech.

North Georgia is a leading coeducational public university emphasizing strong liberal arts, as well as pre-professional,

professional and graduate programs. North Georgia is distinguished by its academic excellence, the quality of its students and graduates, its high graduation rates, and its leadership and military missions. The university's Corps of Cadets has a long tradition of excellence, and, in 2009, North Georgia College & State University was the top-performing senior military college participating the U.S. Army's Leadership Development Assessment Course and in 2010 the Corps of Cadets earned the coveted MacArthur Award for military excellence.

The heart of North Georgia is its historic drill field, and the people who live, study and work around it contribute to a unique experience that educates students for life and leadership for today's global community. In a culture that reflects the school's mission and core values of courage, integrity, loyalty, respect, service, truth, and wisdom – North Georgia is intent on providing an innovative teaching and learning environment, supporting regional development, and educating engaged citizens. The university's unparalleled educational experience prepares students to be professional, civic and military leaders who have the knowledge and skills to address society's most complex issues and pressing needs – locally, regionally and globally.

As part of the university's growing focus on global educational opportunities, North Georgia College & State University has achieved designation as an ROTC Language and Culture Hub, a role that requires the university to offer an array of strategic languages as part of military officer training. These courses, as well as numerous study abroad opportunities, are also available to non-military students in all disciplines.

HISTORY

North Georgia College & State University began as a branch of the Georgia College of Agriculture and Mechanical which was created by the University of Georgia in 1873 from funds from the Morrill Act. William Pierce Price, a local congressman, persuaded officials at UGA to use part of the funds to establish a branch of the newly created college in Dahlonega, GA, Price's birthplace and home. The college opened classes in 1873 with 177 students (98 males and 79 females) making it the first in the state to admit women. Classes were originally held in the old U.S. mint building that was shut down during the Civil War. After the college was awarded the power to grant degrees in 1876, the first graduating class received degrees in 1879. The first graduating class of 4 consisted of 3 men and 1 young women, making it the first institution in the state to award a degree to a female.

The university has always had a military presence since land-grant act schools were required to teach military tactics, but it was not until World War I that the military programs began to grow. The National Defense Act of 1916 that created ROTC also helped establish the military presence that is felt on the campus today. In 1929 the designation "Agricultural" was dropped from the name and the school became North Georgia College. By 1932 the college was reduced to a 2-year junior college. World War II saw a decline in enrollment because so many male students joined the war effort. This changed when an Army Specialized Training Program was placed at the college to train junior officers. After the war the college grew because of young service men using their GI bill to attend school. By 1946 the college was reinstated as a four-year college. In the 1950s Dahlonega provided gold for the leafing of the Georgia Capitol building. It was also at this time that similar efforts to gold leaf Price Memorial Hall began - a project that did not see fruition until the 1970s.

TRADITIONS

• Arch: The North Georgia College Arch, which is located at the campus entrance nearest to Dahlonega's square, was built by the Class of 1951 to commemorate their classmates who died in the Korean War. By tradition, freshmen are not supposed to walk

through the larger arch and instead walk though the smaller arch to the side.
• Bugle Calls:
  - Reveille is played every morning at 7:00 a.m., at which time cadets and civilians alike stop and face the flag.
  - Retreat is played every afternoon at 5:00 p.m., at which time all outdoor activity on campus ceases to pay respect to the American flag. Cadets stand at attention and salute the flag while civilians stop, remove their hats, face the flag, and place their right hand over their heart.
  - Taps is played every evening at midnight (2:00 a.m. on Fridays and Saturdays of open weekends) to indicate the end of the day. Cadets are required to be in their dorms at this time.
• Drill Field: The William J. Livsey Field is located in the heart of the main campus. This field is the parade grounds for the Corps of Cadets and is used for drill and ceremonies. It is also used for recreational activities, though the activities of the Corps take precedence. Students do not cut through the field as a shortcut; instead, they walk around. On April 18, 2009, the drill field was dedicated to retired General William J. Livsey.
• Memorial Wall: The Memorial Wall, located in front of Memorial Hall, was built in 1983 and honors North Georgia students and alumni who died either in service to their country or while attending the university. Students do not enter the area around the wall unless they are stopping to show honor to those listed on the wall.
• Retreat Triangle: The triangle is located near the Drill Field, Dining Hall, and Dunlap Hall. It holds the original retreat cannon, a 1902 three-inch gun, which was fired daily for more than 50 years. The cannon was recently restored by the NGCSU Parents Association. Students do not walk on the triangle or tamper with the cannon.
• Dining Hall Steps: Freshman Cadets are not allowed to use the stairs located just outside of the Dining Hall and must instead walk around on the sidewalk and road to get to the Dining Hall.

THE BOAR'S HEAD BRIGADE

The Boar's Head Brigade is the official designation of the ROTC unit at North Georgia College & State University. It is formed along standard military units at a smaller scale. There are two battalions, and each battalion has three companies. There is one detached Headquarters Company. 1st Battalion is housed in Sirmons Hall and is made up of Alpha, Bravo, and Charlie Companies. 2nd Battalion is housed in Gaillard Hall and is made up of Delta, Echo, and Foxtrot Companies. HHC Company resides in Sanford Hall as does the Assassin detachment, which was formed in 2008. Headquarters and Headquarters Company consists of the Brigade Staff, the Golden Eagle Band, the Blue Ridge Rifles (a nationally recognized rifle drill team), the Nurse Detachment, Color Guard, and the Ranger Challenge Team.

A special company known as Golf Company is sometimes formed during the summer term when Corps participation is low.

The female membership of the Corps moved to Sirmons Hall in 1991 (formerly housed in the Lewis Residence Hall), making Sirmons Hall the first co-ed housing facility on campus, though access to the female residences was restricted by card key. However, co-ed showers and latrines with a changeable sign are used. Now, all Military Resident students live in co-ed dorms: Sanford, Gaillard, and Sirmons Halls.

Military life on campus is full-time during the week. A typical week on campus begins with a full brigade drill on Monday. Military drill starts the academic year with brigade drills in the fall, company drills in the winter, and squad drills in the spring. Companies submit to a "white collar" inspection of quarters each academic term. Quarters are meticulously cleaned, and bunks must meet specific inspection requirements with a "white collar" turn down.

Daily life begins with First Call at 6:45 a.m. and Reveille at 7:00 a.m. Retreat is sounded each day at 5:00 p.m. and Taps is played at 12:00 a.m. For Retreat, everyone on campus stops what they are doing, just like on a military post, and pays respect while the flag is lowered.

Cadet uniform requirements vary by weekday: Monday through Thursday is Army Combat Uniform (ACU), and Friday, Saturday and Sunday is Class "B"s. On some Fridays, under the guidance of the Brigade Commander, cadets are authorized to wear "Campus Casual" attire that consists of a blue NGCSU polo shirt, khaki slacks, dress shoes, and a belt.

Members of the Corps may dress down to "civies" at 5:00 p.m. if academic minimums are met. Otherwise cadets must report for "quarters" (required study hall lasting Sunday through Thursday each week) at 8:00 p.m. There is a quarter's break at 10:30 pm and quarter's taps at 11:00 p.m. NCOs within each company alternate as CQ (Charge of Quarters) monitoring each company hall who stay on duty from 8:00 p.m. until official taps at 12:00 a.m.

Inspections and physical fitness are two things the Boar's Head Brigade prides itself on and are thus alternated during the week beginning with First Call. Monday, Wednesday, and Friday are designated as "PT" days, while Tuesday and Thursday are the "inspection" days. On inspection days, the Corps also has Cadet Professional Development classes (CPD). These classes consist of policies, traditions, and techniques that can be used to foster a successful future at NGCSU and in the Corps. Cadets are required to pass the Army Physical Fitness Test each academic quarter.

The Aggressor Platoon as the OPFOR

The Aggressor Platoon is a co-curricular organization sponsored by the North Georgia College & State University Military Department and is open to all students, both military and non-military. The unit functions as one of ten military organizations and seeks to train students in light infantry, Ranger, Special Operations, and guerrilla warfare tactics. Using these skills, the Aggressor Platoon provides a realistic opposing force (OPFOR) for Pre-Camp and Corps of Cadets field training exercises (FTX).

Members use intense training and demanding FTX's to enhance their levels of discipline, leadership, and tactical proficiency.

The Aggressor Platoon was originally formed in 1963 as an affiliate of the Scabbard and Blade in order to provide an opposing force for juniors before they went to LDAC. Members were selected from the sophomore class based on their prior military experience and motivation. The platoon was nicknamed the "Black Tigers" in reference to the all black fatigues worn by the members in addition to their ability to lurk in the shadows and strike ferociously in an instant. The platoon was reorganized in the late 1970's into its current form taking all volunteers from the school. Aggressors can be seen as leaders in the Corps, Military, and civilian sector. Their motto is "Aggressors Lead the Way!"

To become an Aggressor, eligible students must show up at the Aggressor Rock in uniform and complete a physical fitness test, a timed obstacle course, and a 6-mile ruck march. Upon a successful completion of tryouts, the student may have the opportunity to be a candidate in the platoon. To advance in rank, members must complete all assigned tests scoring 80 percent or above and show outstanding leadership in FTXs and regular training events.

BLUE RIDGE RIFLES

Perhaps one of the most highly regarded and respected units on campus is the Blue Ridge Rifles. The BRR get their name from a volunteer rifle unit that was located in Dahlonega, GA, during the Civil War. After the Civil War was over, members of the unit remained in contact. In the 1950s, North Georgia College & State University decided to from a platoon that specialized in rifle drills and showmanship and called the unit the Honor Platoon. Over the years the name of the unit was changed until finally sticking with The Blue Ridge Rifles. They chose this name to pay homage to the original unit.

The BRR is a nationally acclaimed unit that have performed in many drill competitions across the country. They pride themselves on this fact and are often considered to be one of the best drilling units in the country. They frequently compete with other highly esteemed drill units such as Texas A&M and West Point. Their motto is "Blue and Gray All the Way!"

Chaplain Corps

The mission for the chaplains in the Boar's Head Brigade is to help identify problems in the unit, propose solutions, and help commanders maintain a positive command environment. Chaplains will also maintain the capacity to present religious opportunities to the Corps (on a voluntary basis) and provide information on religious activities in the surrounding community. The cadet chaplains also provide morale and motivation for each Cadet company and therefore help to form the backbone of the unit. Each of the 10 chaplains are in positions of responsibility over morale, mental and spiritual health, and motivation in the Corps of Cadets. The motto of the Chaplain Corps is "Storm the gates of hell!"

The Cadet Chaplain Corps was started in the fall of 2000 with only one brigade chaplain. It has now grown to include one chaplain in each company, one chaplain in each battalion, and the brigade chaplain, all volunteers. In 2008, the Boar's Head Brigade Cadet Chaplain Corps was constituted and became an official specialty unit on campus. The Chaplain Corps hosts a variety of events on campus including Corps Bible study and Prayer Breakfast.

Color Guard

The NGCSU Color Guard is without a doubt the single most important specialty unit on campus. It is their job to safeguard and present the colors (American, State, and Boar's Head Brigade) at each and every function of the Corps of Cadets. Their unit is open to any member of the Corps of Cadets that is willing to participate. The Color Guard is also tasked to present any cadets being honored in the many ceremonies that take place throughout the semester at North Georgia.

THE GOLDEN EAGLE BAND

The Golden Eagle Band of North Georgia College & State University is one of the most unique specialty units on campus. Not only is it a military organization, but it is also an academic class. The GEB is the oldest specialty unit on campus and traces its origins to the founding of the university in 1873. It is also the oldest marching band in the state of Georgia. As such, it has a great level of history and tradition. Its members are always willing to accept new ways of accomplishing tasks as one of the most visible ambassadors of the university.

The Golden Eagle Band's mission is to "provide quality musicianship, discipline, and leadership through both military and musical training. We set and maintain the highest standards to represent North Georgia College & State University's Boar's Head Brigade."

Several times each semester, the Corps of Cadets has reviews and functions in which marching is involved. The Golden Eagle Band serves as "the heartbeat" of the Corps at these functions as they maintain the tempo for marches and perform the many bugle calls that are a required part of military processions. The majority of the band's performances are military processions. However, the band has recently integrated Drum Corps International techniques and shows into their regimen.[27] The GEB goes on a tour across the Southeast every Spring semester. This makes for a more thorough and complete collegiate band experience and entertains those who come to see the military reviews and processions.

Even though the GEB is one of several military units on campus, it is in fact open to both cadets and civilian students. Participation is not limited to those with a background in music as all skills needed to perform in a military band are taught by the instructors and cadre. The GEB has two mottoes used to distinguish between new and veteran players: "Talons of Steel" for new members, and "Raise Hell" for veterans.

Mountain Order of Colombo

In 1960, the Order of Colombo Mountain Platoon was conceived by a group of cadets who were interested in forming a unique organization after watching a demonstration performed by the cadre of the U.S. Army Mountain Ranger Camp. After requesting information and training assistance from the ranger camp, Master Sergeant Louis P. Colombo, who was assigned to Camp Merrill, volunteered his time and knowledge. Prior to MSG Colombo's departure, the unit was named in his honor. MSG Colombo died in November, 1995.

The unit is sponsored by the military department to promote interest in military mountaineering and small unit infantry tactics. Members are selected from those cadets who successfully pass a rigorous physical fitness test and tactical skills test. Their mission is to train and develop cadets into potential mountain combat soldiers with emphasis on the subjects of mountaineering, fixed installations, terrain navigation, small unit tactics, hand-to-hand combat, and survival tactics.

Small unit tactics are initially taught in the classroom followed by practical exercises in a field environment. Mountaineering skills are initially taught in the classroom, practiced on the rappelling tower, and then perfected at Mount Yonah. Their motto is "Mountain All the Way!"

About the Atlanta Vietnam Veterans Business Association (AVVBA)

On November 11, 1981, Don Pardue, Joe Harrison, Don Plunkett, and Mike Turner met after work at a local restaurant, then known as Penrod's. It was Veterans' Day, and it was going to be remembered. This was the birth of the Atlanta Vietnam Veteran's Business Association (AVVBA).

The following year, a few more friends gathered at the 57th Fighter Group Restaurant. The group was very small, but the spirit was very apparent. From there a couple years passed, and Mal Garland came up with the idea to memorialize a fellow Atlantan who had heroically perished in the Vietnam War. Garland offered to be our first official president, Walter Stroman volunteered to be our treasurer, and Steve Martin, Esq. donated his services to create required legal documents to give us definition.

The memorial plaques began in 1987 at the Galleria Complex with our first guest speaker, U. S. Senator Max Cleland. The following year, Duke Doubleday single-handedly orchestrated the next ceremony with his guest speaker and former war boss - none other than Maj. Gen. George S. Patton, Jr.

As the membership continued to increase, the memorial services became more defined. Russ Jobson and Don Pardue supervised the 1989 event located at Peachtree Center. Over the next several years, internal volunteers emerged to help organize the plaque ceremonies resulting in the procession of leadership responsibilities. These members included Russ Jobson, Bill Threlkeld, Lanny Franklin, Pat Garland, Ron David, Max Torrence, and Rick Lester.

Our first educational facility memorial came in 1995 at Georgia Tech under the guidance of Jeff Colbath.

From that extremely humble beginning, 23 monuments have been erected throughout the greater Atlanta area. The organization has succeeded in bringing praise, peace and graceful closure to those family members who suffered the loss of these great American heroes. The pride harbored by the AVVBA mission continues to help our members, relatives, and friends of our memorial honorees and all Atlantans focus on the positive aspects that have arisen from that conflict.

Although the annual memorial ceremonies are a significant part of AVVBA's actions, we do much more than that. Monthly lunch meetings, held the first Tuesday of each month, bring 75-100 members together monthly to share fellowship with each other, listen to a guest speaker, and participate in a raffle, proceeds which are donated to the Atlanta Airport USO. In a typical month, a check for $450 - $500 is presented to the USO to support our troops.

Every Veterans Day, AVVBA members march in Atlanta's Veterans Day parade. Unlike the stereotype of the Vietnam veteran, the unit regularly is recognized with parade trophies for their sharp appearance in the distinctive white shirts, AVVBA tie, blue blazers, and khaki pants - and they aren't too shabby a marching unit considering they only march once a year.

Since 2004, AVVBA has been a leading volunteer organization at the Atlanta Airport USO. At least twice each month, eight to twelve volunteers spend a day doing whatever is required to help the USO welcome and take care of America's military troops as they come and go from war zones in Iraq, Afghanistan, and Kuwait. In 2007 in conjunction with the Atlanta chapter of International Brotherhood of Electrical Workers (IBEW), they parked cars in the IBEW parking lot at Atlanta Braves games and raised $40,000 to contribute to the Atlanta USO. AVVBA volunteers manned the parking lot during each of the Braves home games.

AVVBA members regularly volunteer to speak at school, church, and civic group events. Helping to teach patriotism to America's students and other citizens is something these veterans are very good at doing. Our members adhere to the statement from our web page, "To those who fight for it, life has a flavor the protected never know." To invite an AVVBA member to speak to one of your groups, contact us on our web site at www.avvba.org.

Now with over 300 members, AVVBA welcomes new members - Vietnam veterans who adhere to our motto - Proud, Professional, Patriotic.

AVVBA MEMORIALS

Each year since 1987, on the Thursday before Memorial Day, AVVBA has dedicated a memorial honoring someone from the Atlanta area who lost his life in Vietnam. Two NGC students have been honored with a memorial. Pictures of all the memorials can be seen at www.avvba.org.

1987
1st LT John L. Fuller, Jr.
Galleria Complex

1988
1st LT Gary C. Jones
CNN Center Atrium

1989
CWO Robert N. Sauls
NGC Student
Peachtree Center

1990
PFC Joel C Roper
Concourse Office Complex

1991
WO Francis McDowell, Jr.
NGC Student
Hartsfield International Airport

1992
CPL Richard F. Sutter
Underground Atlanta

1993
1st LT William E. Gay, Jr.
Shepherd Spinal Center

1994
CPL Charles H. Brittian, Jr.
Georgia World Congress Center

1995
Major Peter P. Pitman
Georgia Tech

1996
CPT M. Dale Reich, Jr.
First Union Plaza

1997
Major Joseph A. Bishop
Woodward Academy

1998
LCPL Russell M. Dobyns, Jr.
Chastain Park

1999
WO George T. Condrey III
Lenox Towers

2000
All Atlantans KIA in
SouthEast Asia
Atlanta History Center

2001
CPT J. Patrick Jaeger
Two Live Oak Building

2002
SSG Allen B. Callaway
Dekalb County Courthouse

2003
Captain Frank Eugene Fullerton
Harold R. Banke Justice Center
Jonesboro, Georgia

2004
Specialist 4 Michael
Robert Glenn
Smyrna Town Center, Smyrna,
Georgia

2005
LT Commander Clarence
William Stoddard
Colony Square,
Atlanta, Georgia

2006
Major William Henry Seward
Perimeter Place

2007
PFC Jerry Wayne Gentry
Cartersville Visitor Center,
Georgia

2008
Major James Carl Wise, Jr.
East Cobb Park

2009
SGT Preston Tribble , Jr.
Millennium Gate at Atlantic
Station